Design, Manufacturing, and Application of Negative Poisson's Ratio Structure

Guan Zhou · Wanzhong Zhao · Yingxin Hu ·
Chunyan Wang

Design, Manufacturing, and Application of Negative Poisson's Ratio Structure

Springer

Guan Zhou
Nanjing University of Aeronautics
and Astronautics
Nanjing, China

Yingxin Hu
Nanjing University of Aeronautics
and Astronautics
Nanjing, China

Wanzhong Zhao
College of Energy and Power Engineering
Nanjing University of Aeronautics
and Astronautics
Nanjing, Jiangsu, China

Chunyan Wang
College of Energy and Power Engineering
Nanjing University of Aeronautics
and Astronautics
Nanjing, Jiangsu, China

ISBN 978-981-96-1936-8 ISBN 978-981-96-1937-5 (eBook)
https://doi.org/10.1007/978-981-96-1937-5

This Springer imprint is published by the registered company Springer Nature Singapore Pte Ltd.
The registered company address is: 152 Beach Road, #21-01/04 Gateway East, Singapore 189721, Singapore

Contents

1 Overview of Negative Poisson's Ratio Structures 1
1.1 Definition of Negative Poisson's Ratio Structure 1
1.2 History of Negative Poisson's Ratio Structure Development 3
1.3 Negative Poisson's Ratio Structure Classification 7
1.3.1 Porous Material with Negative Poisson's Ratio 7
1.3.2 Negative Poisson's Ratio Polymeric Materials 13
1.3.3 Negative Poisson's Ratio Composites 15
1.4 Special Mechanical Properties of Negative Poisson's Ratio Structures 22
1.5 Study on Mechanical Properties of Negative Poisson's Ratio Structure 29
1.6 Microstructure of Negative Poisson's Ratio Materials 30
1.6.1 Microstructural Deformation Mechanism of Negative Poisson's Ratio Materials 31
1.6.2 Microstructure Design of Negative Poisson's Ratio Materials 33
1.7 Other New Negative Poisson's Ratio Structures and Their Applications 34
1.8 Summary of This Chapter 45
References 45

2 Performance Analysis of Negative Poisson's Ratio Structures 47
2.1 Introduction 47
2.2 Double-Arrow Type Negative Poisson's Ratio Structural Mechanics Model 48
2.2.1 Negative Poisson's Ratio Structure Equivalent Elastic Modulus 50
2.2.2 Negative Poisson's Ratio Structural Shear Modulus 61
2.2.3 Relative Density of Negative Poisson's Ratio Structure 67

2.3 Study of Static Properties of Negative Poisson's Ratio Structure ... 69
2.4 Study of Energy Absorption Characteristics of Negative Poisson's Ratio Structure ... 73
2.4.1 Negative Poisson's Ratio Structure Compression Properties ... 73
2.4.2 Energy Absorption Mechanism of Negative Poisson's Ratio Structure ... 76
2.5 Summary of This Chapter ... 79
References ... 80

3 Design Methods for Negative Poisson's Ratio Structures ... 81
3.1 Introduction ... 81
3.2 Single-Objective Optimal Design of Negative Poisson's Ratio Structure ... 82
3.2.1 Study of Single-Objective Optimization Method and Process for Negative Poisson's Ratio Structure ... 83
3.2.2 Parametric Modeling of Negative Poisson's Ratio Structure ... 88
3.2.3 Stiffness Maximization Design of Negative Poisson's Ratio Structure ... 93
3.2.4 Optimization of Maximum Stress Minimization for Negative Poisson's Ratio Structures Design ... 95
3.2.5 Optimization of Negative Poisson's Ratio Structural Mass Minimization Design ... 97
3.3 Multi-objective Optimization of Negative Poisson's Ratio Structures Design ... 99
3.3.1 NSGA-II Multi-objective Optimization Algorithm ... 99
3.3.2 Multi-objective Optimization of Two-Dimensional Negative Poisson's Ratio Structure Design ... 101
3.3.3 Multi-objective Optimization Design of Three-Dimensional Negative Poisson's Ratio Structure ... 106
3.4 Robustness Optimization of Negative Poisson's Ratio Structure ... 109
3.5 Gradient Optimization of Negative Poisson's Ratio Structure ... 111
3.5.1 Thickness Gradient Type Negative Poisson's Ratio Structure and Design Parameters Selection ... 112
3.5.2 Negative Poisson's Ratio Structural Constraints and Loading ... 114
3.5.3 Optimization Results for the Thickness Trapezoid of the Negative Poisson's Ratio Structure ... 116
3.6 Summary of This Chapter ... 118
References ... 119

4 Manufacturing Method of Structures with Negative Poisson's Ratio 121
4.1 Frontiers 121
4.2 Fabrication of Porous Negative Poisson's Ratio Materials 122
4.2.1 Fabrication of Negative Poisson's Ratio Foam 122
4.2.2 Fabrication of Honeycomb Structure with Negative Poisson's Ratio 123
4.3 Fabrication of Polymer Materials with Negative Poisson's Ratio 130
4.4 Fabrication of Negative Poisson's Ratio Composites 131
4.5 Negative Poisson's Ratio Measurement of Poisson's Ratio of Materials 132
4.5.1 Digital Image Method 133
4.5.2 Tensile Test Stand 135
4.6 Summary of This Chapter 137
References 138

5 Engineering Applications of Negative Poisson's Ratio Structures 139
5.1 Introduction 139
5.2 Application of Negative Poisson's Ratio Structure in Energy Absorption 140
5.2.1 Application of Negative Poisson's Ratio Structure in Automobile Energy Absorbing Box 141
5.2.2 Application of Negative Poisson's Ratio Structure in Anti-collision Bar of Automobile Side Collision 174
5.2.3 Application of Negative Poisson's Ratio Structure in Energy-Absorbing Structure for Pedestrian Protection 187
5.3 Application of Negative Poisson's Ratio Structure in Other Industries 200
5.4 Summary of This Chapter 205
References 206

Chapter 1
Overview of Negative Poisson's Ratio Structures

1.1 Definition of Negative Poisson's Ratio Structure

Poisson's ratio was first discovered and proposed by the famous French mathematician and physicist Simeon-Denis Poisson. It is defined as the ratio of the negative transverse strain to the axial strain when a material is subjected to axial compression or tension. The Poisson's ratio is a fundamental material constant, referring to the ratio of the lateral strain to the longitudinal strain caused by uniformly distributed longitudinal stress within the elastic proportional limit of the material, reflecting the basic deformation of the material. The Poisson's ratio primarily reflects the volume change of the material during tension or compression. The smaller the volume change of the material during tension or compression deformation, the larger its Poisson's ratio. Typically, materials expand laterally when subjected to axial compression or contract laterally when stretched, resulting in a positive Poisson's ratio [1]. It is generally believed that almost all materials, such as steel, copper, rubber, and polymers, have a positive Poisson's ratio.

Most materials typically contract in cross-section when subjected to uniaxial tension from external forces, and expand in cross-section when subjected to uniaxial compression in practical engineering. According to the definition of Poisson's ratio, the Poisson's ratio of a material is the ratio of the lateral strain to the longitudinal strain during uniaxial compression (or tension). As shown in the following equation.

$$\nu = -\frac{\varepsilon_y}{\varepsilon_x} \tag{1.1}$$

From the definition of Poisson's ratio, it is evident that the Poisson's ratio is an important parameter reflecting the properties of a material. The smaller the Poisson's ratio, the greater the lateral strain of the material during tension; conversely, it is smaller. Ordinary materials exhibit a necking phenomenon when subjected to longitudinal stress, and their Poisson's ratio is positive. Materials with a negative Poisson's

G. Zhou et al., *Design, Manufacturing, and Application of Negative Poisson's Ratio Structure*, https://doi.org/10.1007/978-981-96-1937-5_1

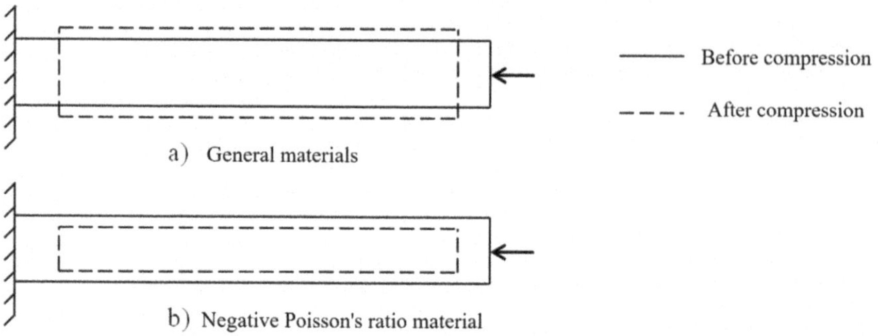

Fig. 1.1 Deformation diagram of general materials and negative Poisson's ratio materials under uniaxial compression

ratio, also known as auxetic materials (derived from the Greek word), expand laterally when stretched and contract when compressed. The Poisson's ratio of a material determines the direction and extent of the change in its lateral cross-section when subjected to external tensile or compressive forces. Generally, materials exhibit positive lateral strain (negative) when compressed (stretched) and negative longitudinal strain (positive), indicating a positive Poisson's ratio. Most materials in nature have a Poisson's ratio between 0 and 0.5. In contrast to ordinary materials, materials with a negative Poisson's ratio contract in their lateral cross-section when subjected to uniaxial compression. Therefore, both the lateral and longitudinal strains are negative. From the definition of Poisson's ratio, it is evident that the Poisson's ratio of such materials is negative. Figure 1.1 shows the deformation diagram of a general material and a negative Poisson's ratio material.

The special deformation mode possessed by negative Poisson's ratio materials makes them different from general conventional materials in terms of mechanical properties, etc. In isotropic materials, the relationship between the different mechanical properties of the materials is as follows.

$$E = 2G(1 + \nu) = 3K(1 - 2\nu) \tag{1.2}$$

$$G = \frac{E}{2(1 + \nu)} = \frac{3K(1 - 2\nu)}{2(1 + \nu)} \tag{1.3}$$

E is the Young's modulus of the material, K is the bulk modulus, ν is the Poisson's ratio of the material, and G is the shear modulus. From Eq. (1.2), it can be seen that for general materials the Young's modulus is generally greater than twice the shear modulus because the Poisson's ratio is positive. However, when the Poisson's ratio of the material is negative, the relationship between the two will change, when the Poisson's ratio of the material is −0.5, Young's modulus and shear modulus are equal, the material will be more difficult to shear deformation and more likely to compressive deformation. In general, when the Poisson's ratio is close to 0.5, the material's shear modulus will be much smaller than the bulk modulus, the material is easy to shear deformation and difficult to produce compression deformation. For negative Poisson's ratio material, when the Poisson's ratio is close to -1, from the Eq. (1.3), the shear modulus of the material will be much larger than the bulk modulus, so the material will be very difficult to occur shear deformation.

1.2 History of Negative Poisson's Ratio Structure Development

Negative Poisson's ratio structures and materials exhibit a unique behavior wherein they expand perpendicularly to the applied force during axial tension and contract in the same direction during axial compression. This extraordinary property not only differentiates them from conventional materials but also enhances their mechanical properties, including shear modulus, indentation resistance, and fracture toughness. Initially discovered in natural substances like certain metals and animal skin, the exploration and production of diverse negative Poisson's ratio materials has propelled the advancement of research in this field, albeit still in its early stages compared to the extensive study and widespread use of other materials [2]. In terms of understanding negative Poisson's ratio materials, novel material theories, mechanical models, and experimental methods need to be employed in the research process. In terms of performance analysis, a comprehensive exploration of dynamic and static response, stiffness, energy absorption properties, and the intricate relationship between performance and material microtopology is imperative. Although the negative Poisson's ratio structure and its exceptional mechanical properties open up a vast realm of potential applications, many of these possibilities remain hypothetical. Consequently, further investigation is warranted to substantiate their viability for large-scale implementation.

People discovered long ago that certain materials, such as the skin around a cow's teat, exhibit negative Poisson's ratio characteristics. These materials, due to their unique properties, have garnered attention from scholars. However, natural materials with a negative Poisson's ratio are rare. The earliest man-made material with this characteristic was obtained in a lab by Lakes from LOWA University in 1987 [3]. He placed regular polyurethane foam into an aluminum mold, subjected it to three-dimensional compression, heating, cooling, and relaxation processes, resulting in a material with a concave unit structure and a Poisson's ratio of -0.7. This research propelled the study of Poisson's ratio materials into a rapid development phase. In 1989, Evans et al. discovered that polytetrafluoroethylene exhibited a negative Poisson's ratio characteristic when the arterial tube made from this material was axially stretched, causing lateral expansion [4]. In 1992, Alderson and Evans found that ultra-high molecular weight polyethylene, produced using a special method, also had this characteristic [5]. Anselm Griffin used liquid crystal polymers to mimic fiber arrangement to study the material's negative Poisson's ratio characteristic [6].

Jeff Moore added metal to foam plastic and found that the resulting metal foam had a negative Poisson's ratio [7]. Warren, Gibson, et al. studied the microstructure of materials with a negative Poisson's ratio, exploring the mechanisms behind this characteristic [8]. Rothenburg studied a macro structure made of rods, sliders, and springs and found that if part of its structure was made concave, it would exhibit a negative Poisson's ratio [9]. Zhengdong et al. studied a special microstructure with a negative Poisson's ratio, expanding the research from micro to macro [10]. Choi et al. conducted a more systematic study on the mechanisms behind materials with a negative Poisson's ratio effect, delving deep into the formation methods of the microstructure and its deformation under external forces [11].

When analyzing anisotropic fibrous compounds, researchers found that some composite exhibit negative Poisson's ratio characteristics. While studying multi-layered fibrous materials, Herakovich believed that by layering in a specific sequence across different scales, a material with a Poisson's ratio value of -0.81 could be achieved [12]. In 1992, Milton successfully synthesized isotropic negative Poisson's ratio composite in two or three dimensions through hierarchical structures. By selecting appropriate components, the negative Poisson's ratio of the material could be less than -1, greatly advancing the development of such materials [13]. Theocaris, using digital homogenization theory to analyze fiber-reinforced composite, found that when they contain star-shaped or concave angular microstructural units, the material exhibits a negative Poisson's ratio [14]. British material scientists B. D. Caddoek and K. E. Evans discovered that specially processed porous polytetrafluoroethylene resin exhibits expansibility, with a Poisson's ratio reaching -12. Due to its complex fibrous nodal structure, it displays an isotropic negative Poisson's ratio as low as -12. The key feature is the "fiber-bundle-node" network structure, which replaces the honeycomb and foam inverted structures. In terms of deformation mechanisms, the nodes undergo translational transformations through fiber movement, resulting in an expansion effect. Wei Gaoyuan and Edwards believe that for composite to have a negative Poisson's ratio, the volume fraction of the dispersed phase must exceed a critical value, and the ratio of the Young's modulus of the dispersed phase to the basis material must be within a certain range [15]. Stagni believes that when the volume fraction of hollow fibers in composite reaches a certain level, the material will exhibit a negative Poisson's ratio [16]. Currently, it is generally believed that there are two ways to give composite a negative Poisson's ratio. The first is to choose an appropriate laying method in the composite material, requiring the laying material to be highly anisotropic. The second is to add negative Poisson's ratio reinforcing fibers to the composite material, delaying fiber pull-out when the material is subjected to tensile force, and causing lateral expansion during pull-out.

Research has found that some metal materials and polymers exhibit negative Poisson's ratio characteristics originating from the molecular scale. These materials mainly include silica crystals, zeolites, and certain metals. To prepare high-stiffness and strength negative Poisson's ratio polymers, the study of molecular-scale negative Poisson's ratio materials has attracted widespread interest from scholars. Evans designed negative Poisson's ratio materials with concave microstructural units at the molecular scale [17]. Browick's research found that a two-dimensional crystalline membrane, similar to a fishing net pattern, has a negative Poisson's ratio characteristic with a Poisson's ratio of −0.37 [18]. Researchers studying zeolite found that the zeolite subunit silicon dioxide exhibits a negative Poisson's ratio characteristic when rotating and separating, later discovering that the Poisson's ratio of α-quartz can reach a minimum of −1. Baughman's research revealed that 69% of cubic crystal elements exhibit negative Poisson's ratio characteristics in the crystal plane direction [19]. Rovat, using orthogonal crystal elasticity theory to study negative Poisson's ratio materials, found that metals like aluminum and copper exhibit negative Poisson's ratio characteristics when the crystal plane angle is between 0 and $\pi/2$ [20]. Compared to ordinary materials, molecular negative Poisson's ratio materials, due to their superior mechanical properties, have broad application prospects in industries such as automotive, shipbuilding, and defense.

Porous Negative Poisson's Ratio structures are an important branch in negative Poisson's Ratio structures and materials, which mainly includes foam materials and negative Poisson's Ratio structures. Porous materials generally refer to composite composed of one phase as solid and another phase as voids. Porous Negative Poisson's Ratio structures can be both two-dimensional and three-dimensional structures. Currently, people have synthesized and manufactured negative Poisson's ratio three-dimensional foam, micro-porous polymers, and negative Poisson's Ratio two-dimensional/three-dimensional structures. Related research shows that the Porous Negative Poisson's Ratio structures mainly exhibit negative Poisson's ratio characteristics due to the structural deformation in its microscopic structural cells. In 1987, a method that could convert the annular positive Poisson's ratio open-cell thermoplastic (polyurethane) foam into negative Poisson's ratio materials was reported. Lakes and Frills believed that polymer foams with softening points, ductile metal foams, and some thermosetting polymer foams could also be made into negative Poisson's ratio materials, and found that the produced copper foam had a Poisson's ratio value of −0.8. Since Lakes successfully prepared negative Poisson's ratio polyurethane foam in 1987, reports on the preparation of draw-ratio thermoplastic (silicone rubber) and metal (copper) have emerged one after another. Thick foams of several tens of centimeters were also successfully prepared. Regardless of isotropic or anisotropic, negative Poisson's ratio foams can be smoothly made. However, the drawback of negative Poisson's ratio foam materials is that the material strength and hardness are too low. If the strength and hardness are improved through enhancement methods, the negative Poisson's ratio effect will disappear. As mentioned earlier, Lakes first successfully prepared negative Poisson's ratio foam materials with concave structures, and its microscopic structure is shown in Fig. 1.2. This hive inverted hexagon can be seen as a regular hive hexagon with the top and bottom vertices interpolated to form oblique-sided "ribs", the material has a flipped hexagonal structure. When it is subjected to axial stretching, the concave sides of the unit open outward, the cell walls undergo linear expansion, thereby increasing the microscopic unit's area, and the material collides in the direction perpendicular to the external force, and vice versa, thus exhibiting a negative Poisson's ratio effect. Bezazi and Scarpa conducted comparative studies on the static and fatigue properties of traditional foam and negative Poisson's ratio foam through experiments, and found

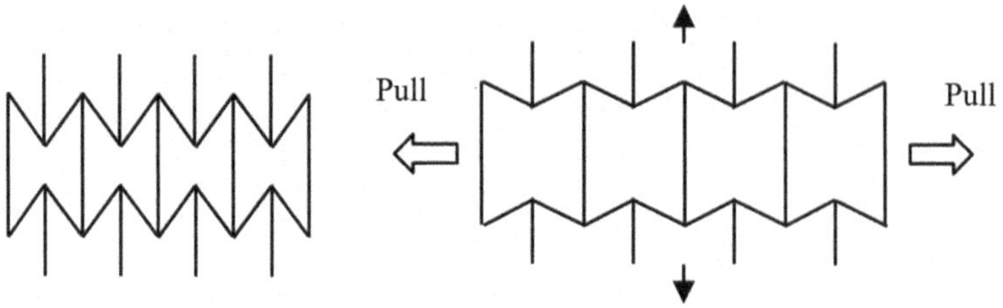

Fig. 1.2 Inner concave structure negative Poisson's ratio foam material

that negative Poisson's ratio foam materials exhibited better mechanical performance during the testing process [21]. Bianchi pointed out through research that negative Poisson's ratio foam materials have shape memory behavior at certain temperatures [22]. Gibson and Ashby pointed out through research that the mechanical properties of porous materials can be jointly described by the geometric parameters of porous structural cells and the mechanical properties of basis materials [23]. The special mechanical properties of negative Poisson's ratio foam materials have broad application prospects in the aerospace and automotive industries as core and filling materials, etc.

The two-dimensional concave porous structure is the earliest discovered honeycomb structure with negative Poisson's ratio effects. When the microscopic units are stretched, the oblique cell walls exhibit movement in the vertical direction, thereby displaying the negative Poisson's ratio characteristic. Figure 1.3a, b represent the chiral and star-shaped two-dimensional porous negative Poisson's ratio structures respectively. The three-dimensional concave porous structures were also found to exhibit negative Poisson's ratio characteristics by researchers subsequently. Lakes and Witt successfully pointed out the deformation behavior of three-dimensional concave units [24] by transforming traditional three-dimensional structures. With the deepening of research on porous negative Poisson's ratio structures, more and more porous negative Poisson's ratio structures with novel microscopic structures are continuously emerging. Searching for and researching microscopic topological structures with superior negative Poisson's ratio characteristics has become a research hotspot in this field.

Due to the special mechanical properties of negative Poisson's ratio structure under the action of tensile bending or pressure, it has a wide range of application prospects, and in recent years has become a research hotspot for universities and related enterprises, and the research of negative Poisson's ratio structure has also made great development. However, compared with other new engineering materials, the research and application of negative Poisson's ratio structure is still in the initial stage, and there is still a lot of work to be done in terms of structural morphology, behavior mechanism, research methods and practical applications of negative Poisson's ratio materials.

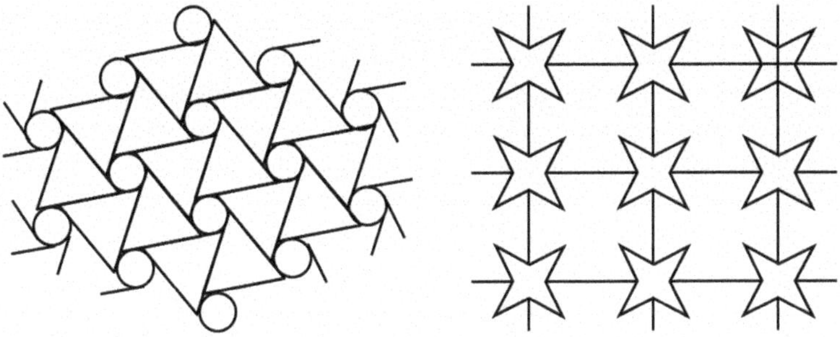

(a)Negative Poisson's Ratio Structure with Hand letter Type

(b)Star type negative Poisson's ratio structure

Fig. 1.3 Porous negative Poisson's ratio honeycomb structure

1.3 Negative Poisson's Ratio Structure Classification

Over the past several decades, scholars both domestically and abroad have discovered and manufactured a large amount of negative Poisson's ratio materials and microstructures through research, including composite, ceramics, metals, and polymers, ranging from the molecular level to the microscopic scale and then to the macroscopic scale. Although, as previously mentioned, there indeed exist negative Poisson's ratio materials with special molecular structures in nature, most of the negative Poisson's ratio materials are still manufactured and synthesized artificially. Due to their inherent unique mechanical properties, negative Poisson's ratio structures and materials have now become a key research area for universities and enterprises, emerging as a new generation of materials with great development and application prospects. Based on the current research on negative Poisson's ratio structures and materials, the commonly seen negative Poisson's ratio materials mainly have the following types.

1.3.1 Porous Material with Negative Poisson's Ratio

Porous materials refer to composite where one phase is solid, and the other phase is entirely composed of voids or liquid, such as rocks and wood found in nature. Porous solid materials may have two-dimensional structures or three-dimensional structures. It has been found that two-dimensional honeycomb solid materials composed of concave pored structural units exhibit negative Poisson's ratio values. Almgren discovered that on a two-dimensional or three-dimensional scale, by designing macro structures composed of rods, springs, and sliders, negative Poisson's ratio effects can be achieved, and its internal structural units are similar to the concave pored units of honeycomb materials. The negative Poisson's ratio polyurethane foam prepared by Lakes in 1987 through thermomechanical methods applied to ordinary foam also belongs to porous negative Poisson's ratio materials. Lakes and Friis, etc., believed that polymer foams with softening points, ductile metal foams, and some thermosetting polymer foams can also be made into negative Poisson's ratio materials, and found that the produced copper foam has a Poisson's ratio value of −0.8. However, the drawback of negative Poisson's ratio foam materials is that the material strength and hardness are too low. If the strength and hardness are improved through enhancement methods, the negative Poisson's ratio effect will disappear.

The above-mentioned materials are all isotropic. For isotropic materials, classical elasticity theory assumes $1 \leq \nu \leq 1/2$. In recent years, negative Poisson's ratio effects have also been found in anisotropic materials such as micro-porous polymers prepared by Evans and others. Caddock and Evans found that some micro-porous materials obtained through special processing methods are anisotropic, having negative Poisson's ratio effects in certain directions, and their Poisson's ratio values are lower than isotropic negative Poisson's ratio materials. Porous negative Poisson's ratio materials generally include honeycomb structures and foam structures.

(1) Honeycomb structure

In the nineteen eighties, scholars first proposed the inner concave hexagonal honeycomb topology in the form of microstructure of dilated honeycomb structure, as shown in Fig. 1.4.

As shown in Fig. 1.4, while subjected to unidirectional tensile load, the concave hexagonal honeycomb structure also shows a tendency to expand in the non-loading direction due to the interaction between the ribs, reflecting the unique tensile-expansion property of negative Poisson's ratio materials. In addition to the inner-concave hexagonal honeycomb structure, many scholars have successively designed honeycomb microstructures with negative Poisson's ratio properties, and several common tensile expansion honeycomb microstructures are given in Fig. 1.5. These structures include inner-concave honeycomb structures, star structures, arrowhead structures, and have both two-dimensional structures and can be expanded into three-dimensional inner-concave honeycomb structures and three-dimensional arrowhead structures.

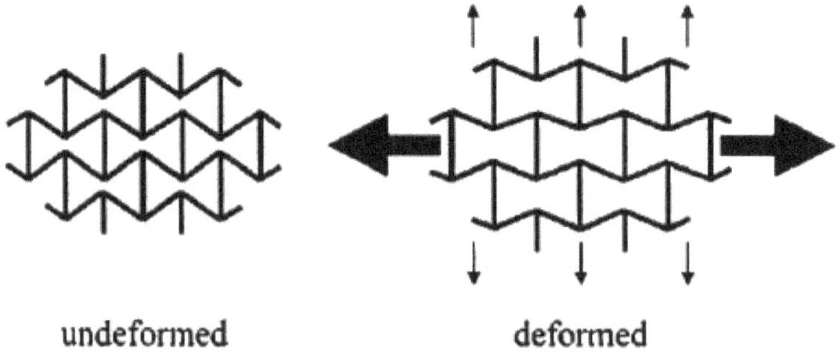

Fig. 1.4 Concave hexagonal honeycomb structure

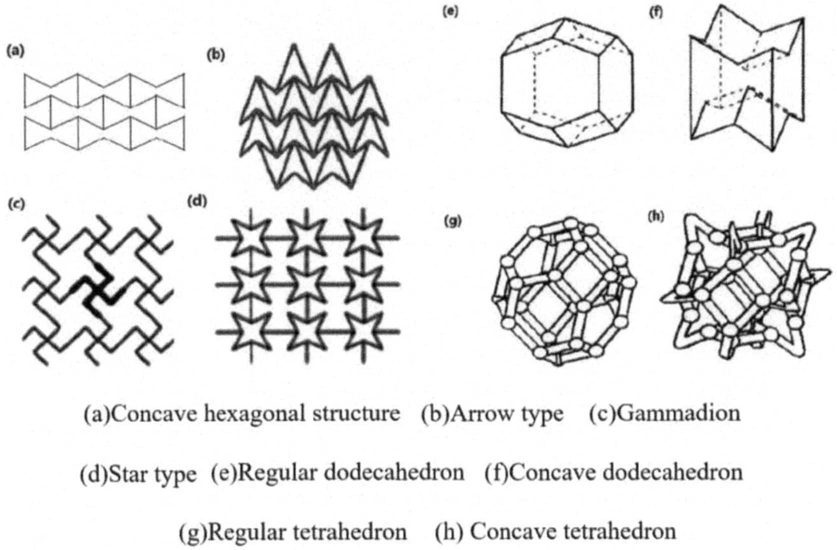

Fig. 1.5 Several common negative Poisson's ratio honeycomb structures

Figure 1.6a shows a chiral structure with negative Poisson's ratio effect through rotational unfolding and contraction of ribs connected by circular cross-sections. Both the star-shaped honeycomb structure in (b) and the arrow-shaped honeycomb structure in (c) provide negative Poisson's ratio effect through bending and rotation of the ribs to control the opening and contraction of the star- and arrow-shaped microstructures.

The chiral system was first proposed by Wojciechowski et al. Later, Lakes et al. developed a six-rod circular node chiral structure, as shown in Table 1.1, which was formed by repeating a mesh structure consisting of circular nodes and six connecting rods tangent to them, in which the angle between adjacent connecting rods was 60°. Prall et al. made the above chiral structure by hand using polystyrene, and through theoretical Prall et al. found that the Poisson's ratio of this chiral structure is around -1 under plane deformation, and its negative Poisson's ratio performance is determined by the shape of the nodes and the length of the connecting rods. In addition to the above six-rod circular node chiral structure, there are also three-rod circular node chiral and anti-chiral structures, four-rod circular node chiral and anti-chiral structures, and four-rod rectangular node chiral structures as shown in Table 1.1, which may also obtain negative Poisson's ratio performance.

Inspired by the tension-expansion toy ball, the Buckling ball has 24 independent circular holes and slots, and there is a certain rotational torque at the vertex connection, which can show obvious compression and tension effects under the action of external force and exhibit peculiar tension-expansion characteristics (Fig. 1.7).

Currently, negative Poisson's ratio honeycomb material is mainly used in two fields: one is as a honeycomb core material to be used as the core layer of the sandwich

(a)double arrowhead structure

(b)star honeycomb structure

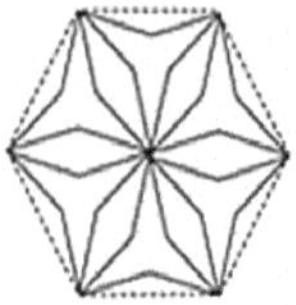

(c)structurally hexagonal re-entrant honeycomb

(d)structure formed from lozenge grids

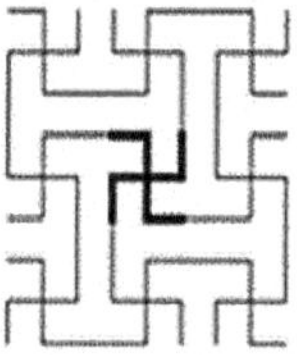

(e) structure formed from square grids

(f) structure formed from sinusoidal ligaments

(g)concave hexagon

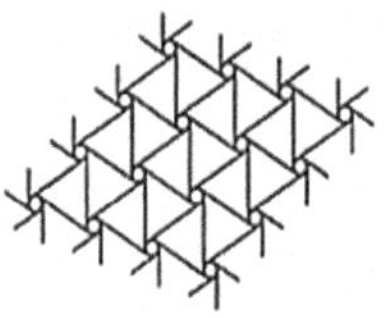

(h)hand type

(i)star type

Fig. 1.6 Some two-dimensional negative Poisson's ratio cellular structure

structure, compared with the traditional honeycomb core, negative Poisson's ratio honeycomb unique hyperbolic properties can be more easily formed into a bending structure, and thus applied in the need to bend the sandwich structure: the other application is as a main material of absorbing and filtering devices, with negative Poisson's ratio effect of the core material will significantly affect the performance of absorbing devices.

(2) Foam structures

In 1987, Lakes compressed ordinary polyurethane foam with a Poisson's ratio of 0.4 into an aluminum mold, then subjected it to heating, cooling, and relaxation treatments, producing polyurethane foam with a Poisson's ratio of 0.7. Since then, the research on negative Poisson's ratio foam structures began to become active. Subsequently, Martz placed low-density polyethylene foam into a specially designed cylindrical sealed chamber, assisted with heating in an oven, applying heat and pressure for a certain period, followed by vacuum maintenance for a while, then reintroduced air

Table 1.1 Chiral structure

Three-rod circular node chiral structure	Four-bar circular node chiral structure	Four-bar rectangular node chiral structure
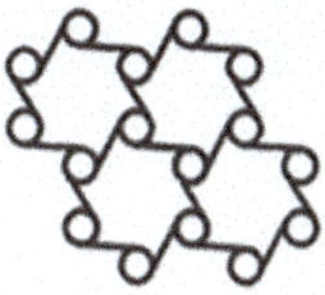	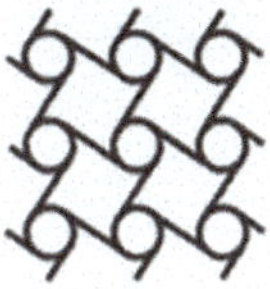	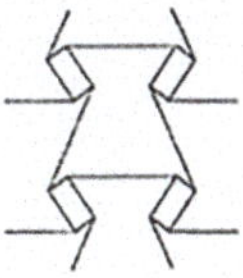
Three-rod round node backhand structure	Four-rod round node backhand structure	Six-rod circular node chiral structure
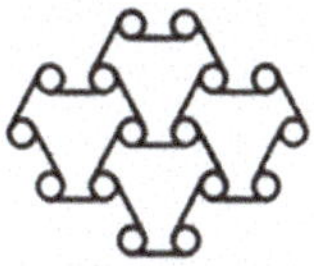	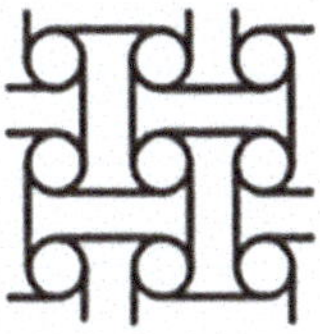	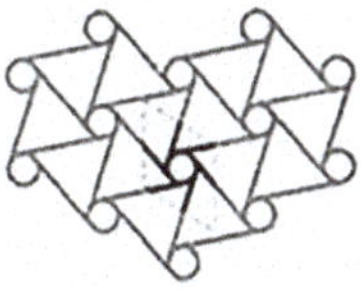

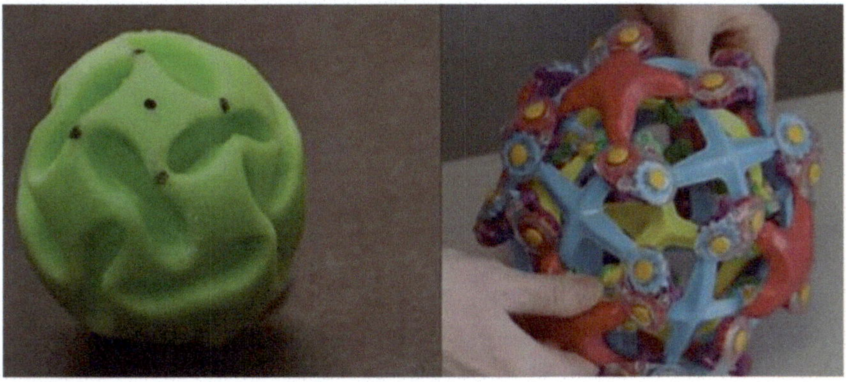

Fig. 1.7 Buckling ball with 24 independent circular hole slots

to return it to normal pressure (i.e., atmospheric pressure), enabling the polyethylene foam to acquire a negative Poisson's ratio concave structure. Through research, it was found that such structures exhibit negative Poisson's ratio properties due to their special concave structure characteristics. Figure 1.8 shows the microscopic structural forms of positive and negative Poisson's ratio foam structures.

Li et al. conducted compression and bending tests on negative Poisson's ratio foams of Ti–6Al–4V alloy with different concave angle structures, showing that the smaller the inner concave angle and the larger the edge length ratio H/L (as shown in Fig. 1.9), the higher the average compression and bending stiffness of the foams and the lower the energy absorption performance of the material.

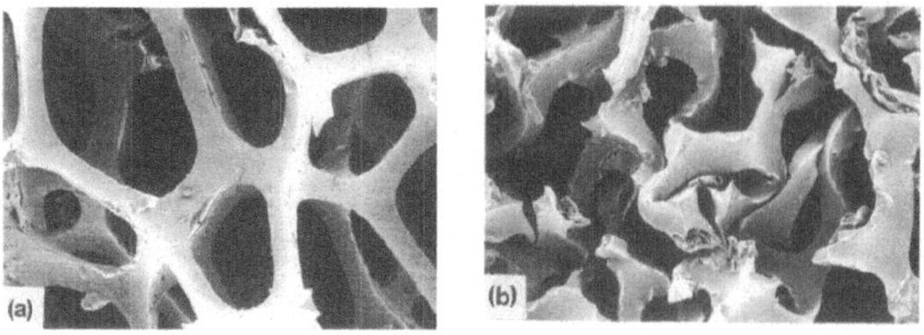

(a)Microstructure form of positive Poisson's ratio foam structure

(b)Microstructure form of negative Poisson's ratio foam structure

Fig. 1.8 Microstructure form of positive and negative Poisson's ratio foam structure

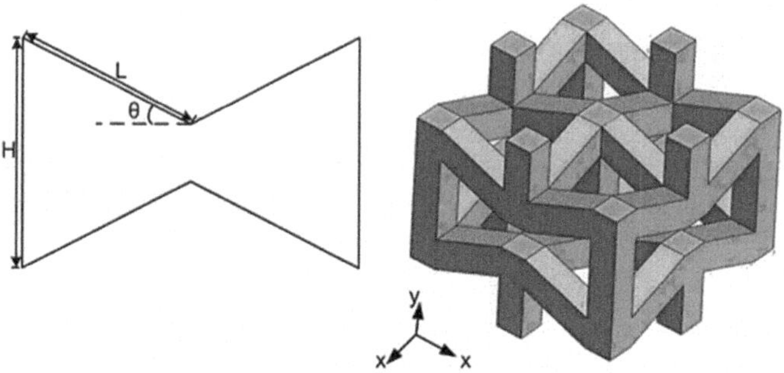

Fig. 1.9 Negative Poisson's ratio Ti–6Al–4V foam structural unit

By adjusting the parameters of foam production process, scholars prepared polyester, metallic copper, polyethylene, polyisocyanine and polyurethane foams with negative Poisson's ratio effect, the essence of which is to prepare a microstructure of concave polyhedral structure with concave angle, such as the 24-sided three-dimensional concave angle structure shown in Fig. 1.10, which is abstracted as a hexahedron with rod unit structure, and the rod units of all six sides are folded and concave to the inside, so that when the foam is stretched, the rod units unfold from inside to outside and expand in space; when the foam is compressed, the rod units continue to fold inward and contract in space.

Since then, copper foams with negative Poisson's ratio properties, silicone rubber foams, and negative Poisson's ratio foam sheets of the order of centimeters in large size have been successfully prepared. The results of a study on the dynamic effects of negative Poisson's ratio foams funded by the U.S. Office of Naval Research showed that negative Poisson's ratio foams have a better ability to absorbing acoustic waves and vibrations than conventional positive Poisson's ratio foams.

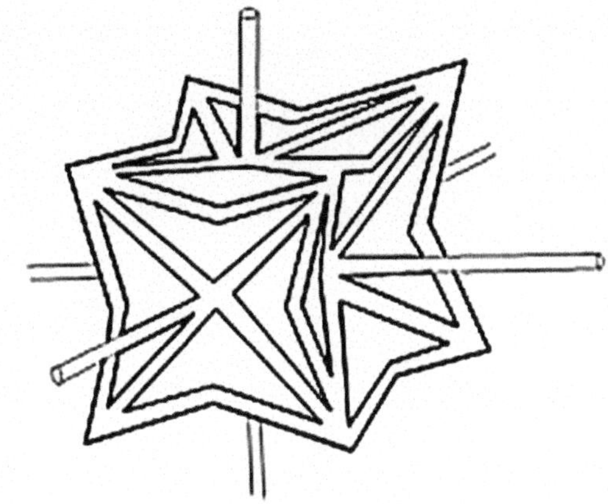

Fig. 1.10 Microstructure of negative Poisson's ratio concave angle

1.3.2 Negative Poisson's Ratio Polymeric Materials

Molecular negative Poisson's ratio materials include certain polymers with special microscopic structures and some crystalline materials, such as zeolites, silicon dioxide crystals, and some elemental metals. It's generally believed that the Poisson's ratio of most polymers is around 1/3. However, some polymer gels, under strict temperature conditions, exhibit negative Poisson's ratio effects when phase transitions occur. As materials with complex microstructures, Evans et al. observed the negative Poisson's ratio effect of microporous polytetrafluoroethylene materials for the first time in 1989. Similar to previous honeycomb and foam structures, the observed polymer materials exhibit negative Poisson's ratio effects due to their specific microstructural configurations. One of the microstructures of negative Poisson's ratio polymer materials is shown in Fig. 1.11 below. Subsequently, Griffin et al. used liquid crystal polymers as the main chain of molecular structures, and through synthetic methods, attached rod-shaped molecular units to the main chain. Such molecular design allows the material to expand laterally under tensile stress conditions, achieving a negative Poisson's ratio effect, which was confirmed through X-ray scattering methods. Bowick et al. discovered through Monte Carlo methods that a two-dimensional crystalline membrane with a microscopic structure similar to a fishnet has a Poisson's ratio value of −0.37.

In 1995, Evans innovated the design of a planar concave corner structure of negative Poisson's ratio molecular material with a honeycomb-like shape as shown in Fig. 1.12, constructing a planar concave corner structure with the benzene ring at the apex of each corner and the carbon atom linked to the benzene ring at the neighboring position to form a pulling and swelling effect.

Wei published a model for calculating the Poisson's ratios of positive and negative Poisson's ratio polymer materials in 1991, introducing centipede-shaped, biped centipede-shaped polymers, and single-arrow, double-arrow shaped small molecules that form stretchable molecular networks through hydrogen bond self-assembly,

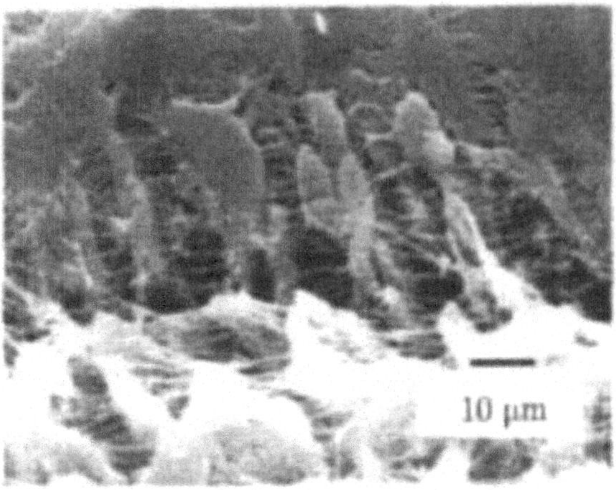

Fig. 1.11 Microstructures of polymeric materials with negative Poisson's ratio

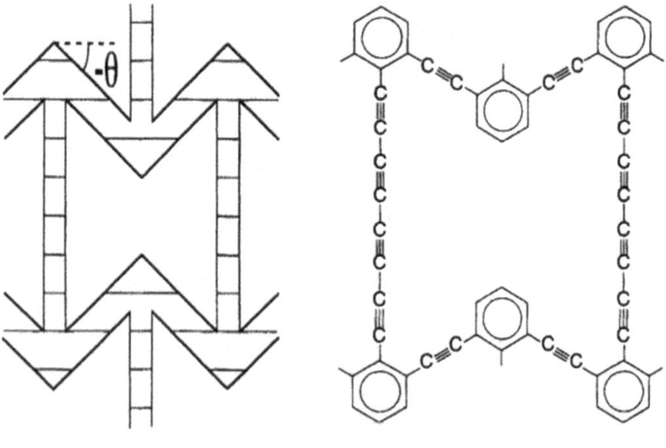

Fig. 1.12 Benzene ring macromolecules with a concave angle structure

illustrating stretchable polymer models. Through numerical analysis, the stretchable conditions for negative Poisson's ratio polymer materials were provided. Subsequent theoretical analysis was conducted on the elastic properties of negative Poisson's ratio ellipsoidal composite. A design for stretchable copolymers consisting of self-assembled arrow-shaped hard-segment and soft-segment alternating molecular networks was proposed. Through molecular mechanics, a metastable supramolecular structure was designed, as shown in Fig. 1.13. Its theoretical negative Poisson's ratio can reach −11.86, and its microscopic structure is also a concave planar concave angular structure.

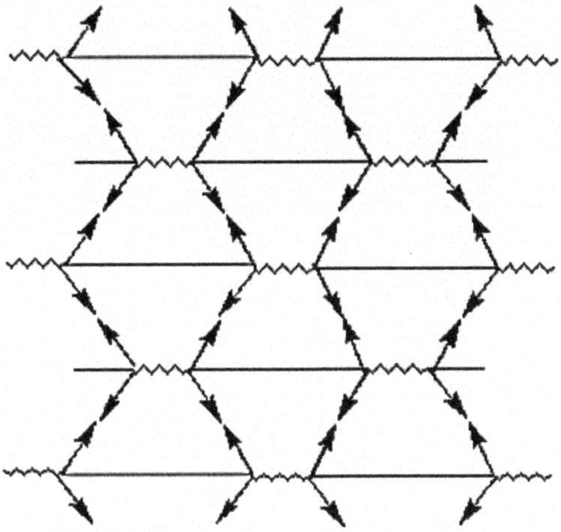

Fig. 1.13 Molecular model of alternating soft and hard block copolymers

Research shows that zeolites with a polyhedral framework microstructure composed of nanoscale molecular chains, due to containing molecular-scale regular voids, exhibit negative Poisson's ratio values. In the study of α-cristobalite it was discovered that the subunits of silicon tetroxide that compose this crystalline material, during their rotation and separation processes, are able to produce negative Poisson's ratio effects. In subsequent research, researchers further calculated the minimum Poisson's ratio of α-cristobalite to reach -1 through rotating tetrahedral model, dilating tetrahedral model, and concurrent tetrahedral model respectively.

For some metallic elements, it was found that they also have negative Poisson's ratios. 69% of the cubic crystalline elements have negative Poisson's ratios in the direction of the crystallographic planes, as indicated by Baughman. Rovat obtained negative Poisson's ratios for copper, aluminum, and nickel alloys (CuAlni) between the crystallographic plane intersection angles of $(0, \pi)$ by simulations using orthogonal crystal elastic symmetry theory.

Due to the higher strength and stiffness of molecular and nanoscale negative Poisson's ratio fibrous polymers, research conducted by many scholars in this field indicates that the negative Poisson's ratio effects of most metallic materials and inorganic materials are caused by special microstructures at the molecular scale. Their molecular-scale concave cell units and concave hexagonal honeycomb structures are similar in form. Figure 1.14 presents several polymer structures with negative Poisson's ratio properties at the molecular scale that have been studied so far.

1.3.3 Negative Poisson's Ratio Composites

Negative Poisson's ratio effects have also been found in some anisotropic fiber-filled composite. For instance, Herakovich discovered that in some multilevel (hierarchical)

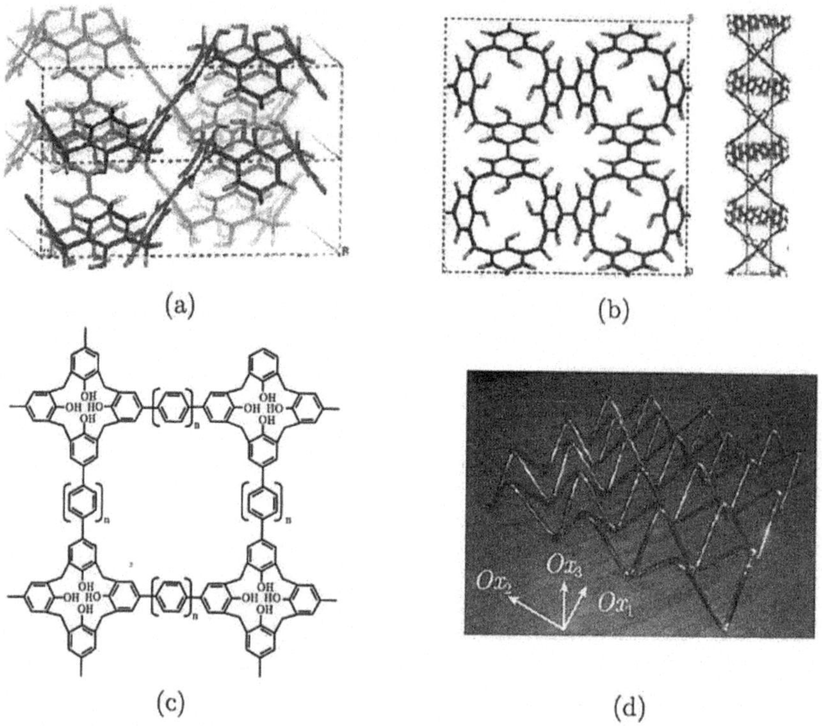

Fig. 1.14 Several polymeric structures with negative Poisson's ratio properties

materials composed of fibers, by controlling the order of layering at different scales, materials with a Poisson's ratio of −0.81 can theoretically be obtained in the direction perpendicular to the fiber layers. Miki and Morotsu also believe that controlling the layering order can endow multilevel materials with negative Poisson's ratio effects, although their research is limited to theoretical analysis.

Evans' research team started to model the highly anisotropic microstructure of polytetrafluoroethylene from microcell concave inward and microcell plus protofibril flexed by connecting microcells in 1989. When stretched along the direction of the protofibril unit, the initially concave microcellular structure (as shown in Fig. 1.15a, b) swells by misaligned lifting or swells by unfolding of the inwardly folded microcellular (as shown in Fig. 1.15c, d). They successfully prepared negative Poisson's ratio polytetrafluoroethylene polymers. The team also reported the fabrication method of negative Poisson's ratio multi-microcellular type ultra-high molecular mass polyethylene in 1992, and discussed its static and dynamic modulus properties for the highest synthetic hardness negative Poisson's ratio material mouth at that time.

In 1992, Milton first prepared multilevel structured negative Poisson's ratio composite with isotropic mechanical properties in two or three dimensions. By

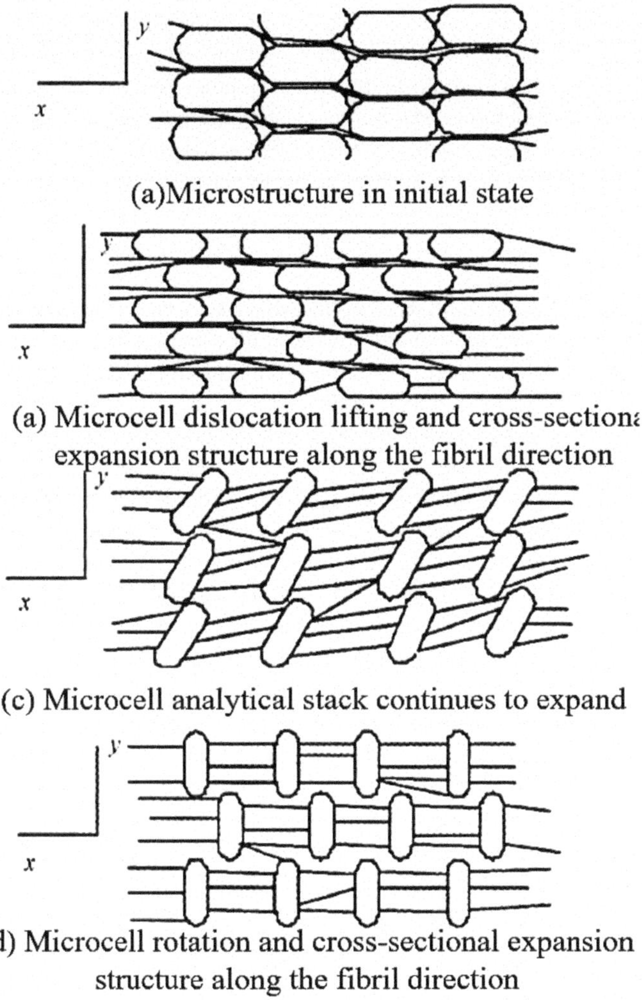

Fig. 1.15 Schematic diagram of PTFE fiber tensile structure changes

controlling the dimensions of each layer and selecting appropriate continuous phase components, he obtained negative Poisson's ratio materials with a Poisson's ratio value close to -1, achieving a leap from theoretical to experimental stage for negative Poisson's ratio composite.

As the theories and models for researching porous solid negative Poisson's ratio materials have become more mature over the years, there has been a gradual increase in research regarding negative Poisson's ratio composite. Theocaris and others, through finite element analysis based on digital homogenization theory, believed that if the internal part of fiber-reinforced composite contains microstructural units with star-shaped cross-sections and concave angles, negative Poisson's ratio effects can be exhibited. Wei Gaoyuan and Edwards theoretically analyzed that when the volume fraction of the dispersed phase in composite exceeds a certain critical value and the ratio of Young's modulus of the dispersed phase to that of the matrix is within a certain range, the materials exhibit negative Poisson's ratio effects. They further concluded through mean field theory, two-dimensional plane, and three-dimensional volume interaction theory that in isotropic spherical inclusions, when the Poisson's ratio value of the matrix decreases from 1/2 to zero, the volume fraction of spherical dispersed phase exhibiting negative Poisson's ratio effects decreases from 0.4174 to zero.

Stagni found that in multilayer hollow fiber-reinforced composite, when the hollow fibers reach a certain volume fraction, they can become negative Poisson's ratio materials and simultaneously increase the effective lateral elastic modulus of the material. In 2009, Victor et al. introduced the concept of negative Poisson's ratio thin-walled structures (as shown in Fig. 1.16), the negative Poisson's ratio properties of polypropylene film are also due to its microscopic foldable structure, which unfolds and continues to fold during stretching or compression. Compared to positive Poisson's ratio structures, negative Poisson's ratio polypropylene fibers and films, when subjected to tensile testing (as shown in Fig. 1.17), exhibit different cross-sectional changes, possessing good load-bearing capacity and energy absorption, larger failure strain, compression buffering ability, smaller density, with a maximum load over 2 times that of traditional fibers, and the energy required for tensile fracture increased by more than 2 times. By adopting an inner spiral folding structure, planar negative Poisson's ratio structures can be constructed, such as rotating rectangles, parallelograms, triangles, and diamonds, by linking them together at vertices in a certain manner to form structural materials with negative Poisson's ratio effects (as shown in Fig. 1.18). When the material is stretched in one direction, the angles between the interconnected rectangles or triangles change. With the increase in angle, the material folds inward perpendicular to the stretching direction, causing expansion deformation, i.e., a stretching effect. Recently, Yang Mingbo found that special polyolefin blends prepared by blending also possess negative Poisson's ratio properties, obtaining a stable negative Poisson's ratio value of -1.4, and the modulus and strength of the material are significantly improved compared to porous negative Poisson's ratio materials.

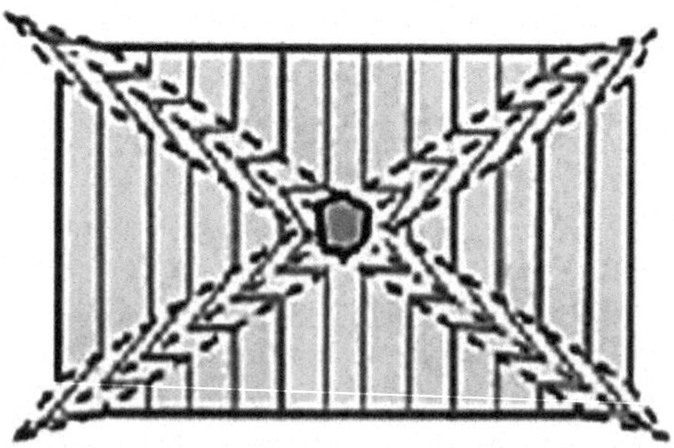

Fig. 1.16 Structure of auxetic film

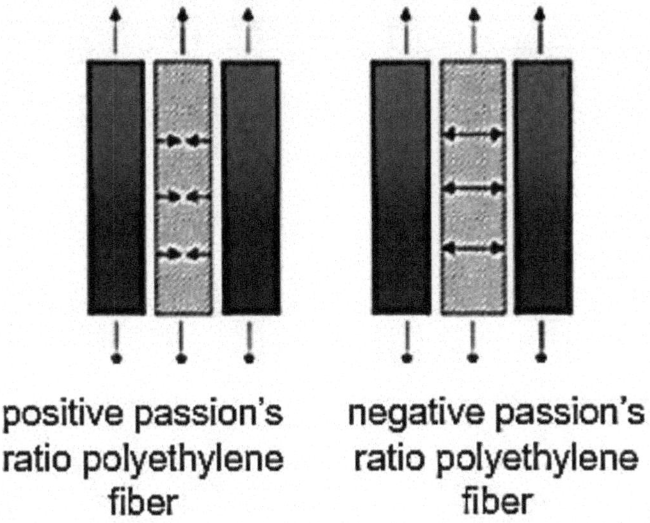

Fig. 1.17 Tensile comparison test for positive and negative Poisson's ratio polypropylene fibers

Miller proposed a new type of negative Poisson's ratio spiral yarn structure, marking the first successful preparation of negative Poisson's ratio fiber composite. This structure is created by winding two types of yarns of different thicknesses and stiffness together in a spiral manner, as shown in Fig. 1.19. Initially, the thicker elastic core is straight while the thinner high-stiffness wrapping is spirally wound around it. At this point, the horizontal length of the structure equals the length of the elastic core. When subjected to tensile loads, both the thin wrapping and the thick core elongate, but the high-stiffness wrapping matches the lateral deformation of the core, extending the length of the structure. The overall lateral diameter of the structure expands to its maximum width, thus exhibiting a negative Poisson's ratio effect.

Ge, through textile technology, introduced a new type of negative Poisson's ratio woven structure, as illustrated in Fig. 1.20. Unlike standard woven structures where the warp and weft yarns are interlaced alternately, in this structure they are arranged alternately and connected through stitching yarns. Under load, the warp yarns assume a wavy shape, thereby generating a negative Poisson's ratio response.

Currently, there are mainly two methods to enable composite to exhibit negative Poisson's ratio effects. The first method is through designing the ideal layering method to achieve an overall negative Poisson's ratio effect. Research results indicate that in order to realize this effect, a type of layering material with a high degree of anisotropy is required, where carbon fiber materials and epoxy resin materials are more suitable. The other method is by incorporating negative Poisson's ratio reinforcing fibers into the material matrix to enable the overall composite material to exhibit negative Poisson's ratio effects.

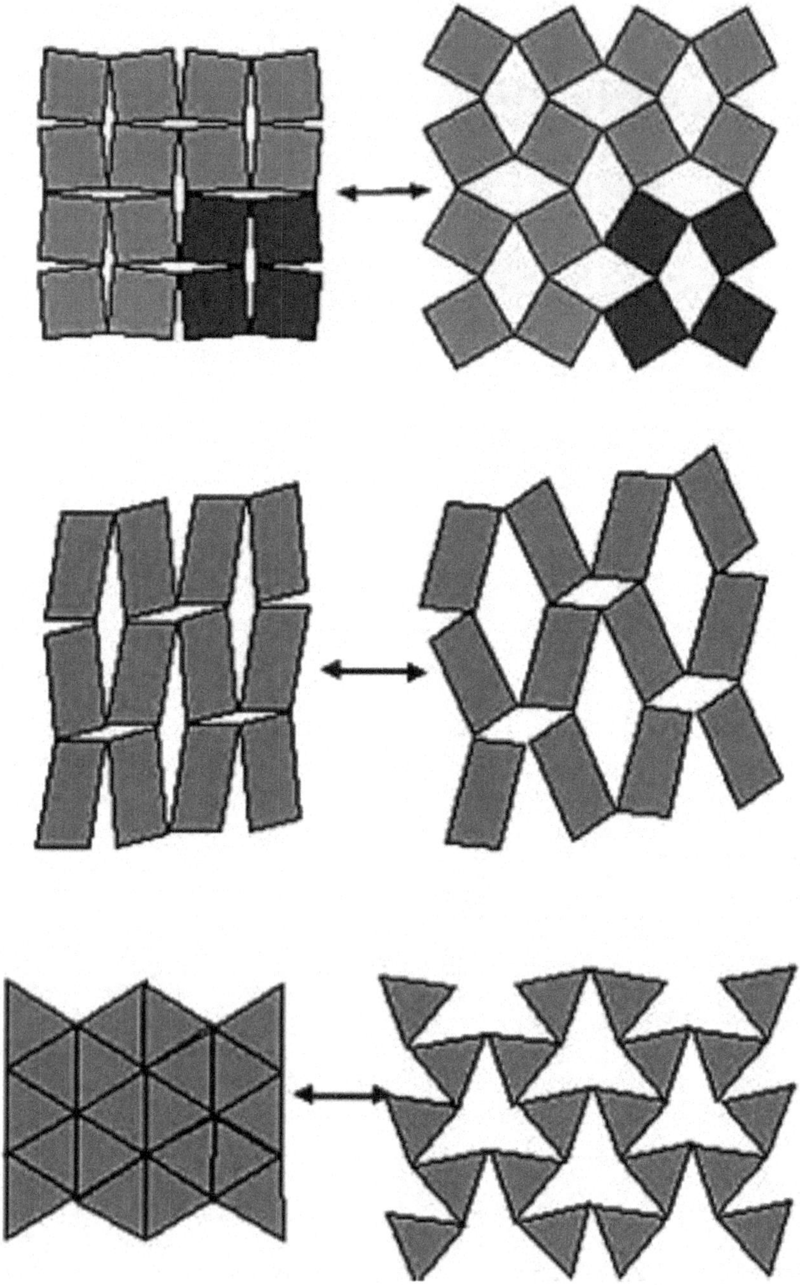

Fig. 1.18 Schematic diagram of the geometric structure of the expanding material of a rotatable structure

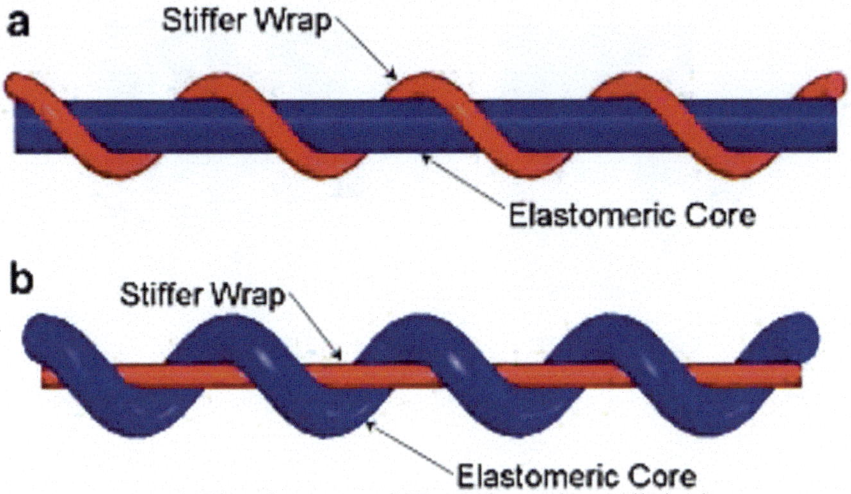

(a)There is no stress (b)Expansion under tensile load

Fig. 1.19 Spiral yarn structure

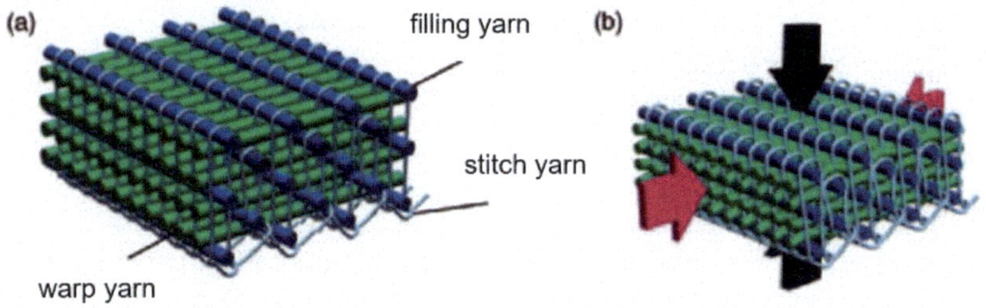

(a)Woven negative Poisson's ratio structure (b)Shrinkage under compressive load

Fig. 1.20 Woven negative Poisson's ratio structure

In the quest for more and better negative Poisson's ratio materials, many scholars use topology optimization technology as a design method, by designing the distribution of materials at the sub-structural level to realize the design of negative Poisson's ratio materials. Utilizing this method can help discover and obtain microstructural forms with specific negative Poisson's ratio characteristics based on actual needs. A typical structure using this method is shown in Fig. 1.21.

With the development of science and technology and the increasing maturity of 3D printing and other advanced manufacturing technologies, more and more complex

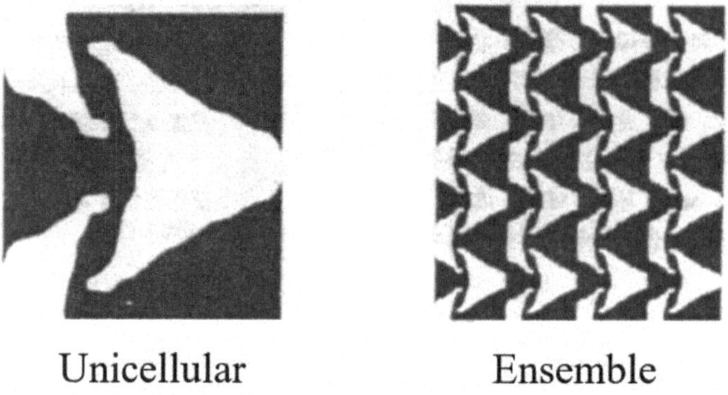

Fig. 1.21 Obtaining a negative Poisson's ratio microstructure using topological optimization

negative Poisson's ratio materials can be prepared directly through 3D printing equipment, which undoubtedly greatly promoted the development of negative Poisson's ratio materials.

1.4 Special Mechanical Properties of Negative Poisson's Ratio Structures

(1) Indentation resistance

With the discovery and preparation of more and more materials with negative Poisson's ratio, how to maximize the discovery and utilization of the potential of negative Poisson's ratio materials has become an important topic. Compared to materials with positive Poisson's ratio, those with negative Poisson's ratio have more unique properties. In a 1999 article by Chan and Evans, it was proposed that under the action of impact loads, traditional materials will compress and flow around from the impacted area in directions perpendicular to the impact load, as shown in Fig. 1.22a. In contrast to traditional materials with positive Poisson's ratio, materials with negative Poisson's ratio have stronger indentation resistance. During the compression process at some part of the material, due to the special deformation properties of the negative Poisson's ratio material, the material around the compressed part will gather towards the compressed part to resist the pressure, as shown in Fig. 1.22. At this time, the local density of the negative Poisson's ratio material increases and the modulus rapidly increases, thus being able to resist the impact more effectively. Research shows that under the same initial density conditions, foams with negative Poisson's ratio have lower stiffness and higher yield strength compared to traditional foam materials. Therefore, it is evident that using negative Poisson's ratio porous materials to make protective devices will yield better impact resistance effects.

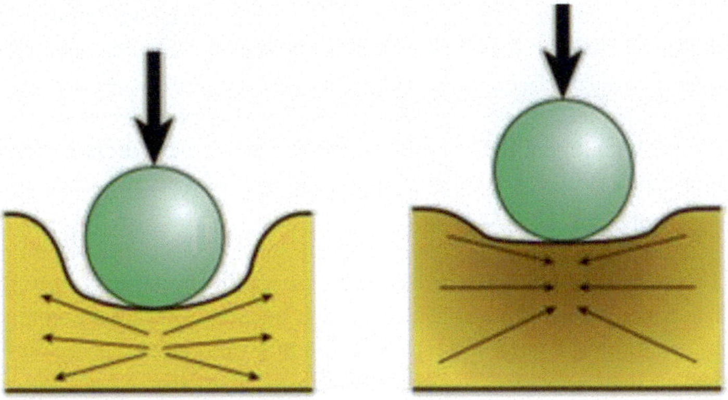

Fig. 1.22 Compression behavior of positive and negative Poisson ratio materials

From the theory of elasticity, it is known that the indentation resistance and hardness of an isotropic material are inversely proportional to $1 - \nu^2$, as shown in the following equation.

$$H \propto \left(\frac{E}{1-\nu^2}\right)^{\gamma} \tag{1.4}$$

$\gamma = 1$ indicates uniform pressure distribution, Herts indentation at $\gamma = 2/3$. From Eq. (1.4), it can be seen that when Young's modulus E is certain, the larger the negative Poisson's ratio ν is, the larger the indentation resistance will be, and the hardness will tend to infinity when $\nu \to -1$.

(2) Shearing Strength

According to the relationship between the modulus of elasticity, shear strength and Poisson's ratio of isotropic materials, it is known that uniform isotropic materials with negative Poisson's ratio have greater shear strength and can better withstand shear forces than positive Poisson's ratio materials.

The elastic properties of a material involve the following parameters: Young's modulus (E), shear modulus (G), bulk modulus (K), and Poisson's ratio (ν). For isotropic materials, the above four parameters are not independent and their interrelationship can be expressed as.

$$G = \frac{E}{2(1+\nu)}, \; K = \frac{E}{3(1-2\nu)} \tag{1.5}$$

When Young's modulus E is fixed, if Poisson's ratio ν decreases from positive to negative, the shear modulus G will be relatively increased, while the bulk modulus K will be relatively reduced, which means that the compressibility of the material will be improved and difficult to shear; if Poisson's ratio can be reduced to -1, G will be relatively K much larger, the material becomes very difficult to shear but easy to volume deformation.

(3) Resistance to cracking

Choi and Lakes in 1996 used experimental and theoretical analysis to propose that negative Poisson's ratio materials have better resistance to cracking compared to conventional materials. When a negative Poisson's ratio material is subjected to tensile action and cracks appear, its cell cells will automatically expand and close the cracks, which indicates that such materials have a stronger resistance to cracking. The tests also demonstrated that the inner concave foam structure exhibited higher fracture toughness than the conventional foam structure at higher body fraction ratios.

(4) Sound absorption characteristics

Haberman mentioned in his 2012 study that auxetic foam materials have better sound absorption properties compared to traditional foam materials, and they perform better in absorbing sound waves below 1500 Hz. In a study concerning ultra-high molecular weight polyethylene materials with negative Poisson's ratio, it was found that the sound absorption capability of auxetic foam materials is 1.5 times that of sintered foam and 3 times that of traditional foam.

(5) Shape memory capacity

Bianchi et al. found through their study that negative Poisson's ratio foam has shape memory ability in a specific temperature environment, i.e., negative Poisson's ratio foam can be restored to its initial size and has positive Poisson's ratio, and the above shape memory behavior can be repeated when the foam restored to its initial state is reconstituted as negative Poisson's ratio foam.

(6) Loss factor (energy absorption)

Foam materials with negative Poisson's ratio have a higher loss factor compared to foam materials with positive Poisson's ratio, hence they hold great potential for applications in energy absorption devices and viscoelastic damping.

Negative Poisson's ratio materials exhibit superior advantages in energy absorption, such as the absorption of ultrasonic waves, acoustic waves, and damping. Through empirical testing, Scarpa et al. demonstrated that auxetic foam has better sound absorption capabilities at lower frequencies compared to traditional foam. Through experimental comparisons between directly prepared ordinary foam and secondarily prepared ordinary foam (converted from auxetic foam), as well as directly prepared auxetic foam and secondarily prepared auxetic foam (converted from ordinary foam), Bianchi et al. found that both directly and secondarily prepared ordinary foam have higher energy absorption rates when stretched, while the secondarily prepared auxetic foam has the highest energy absorption efficiency.

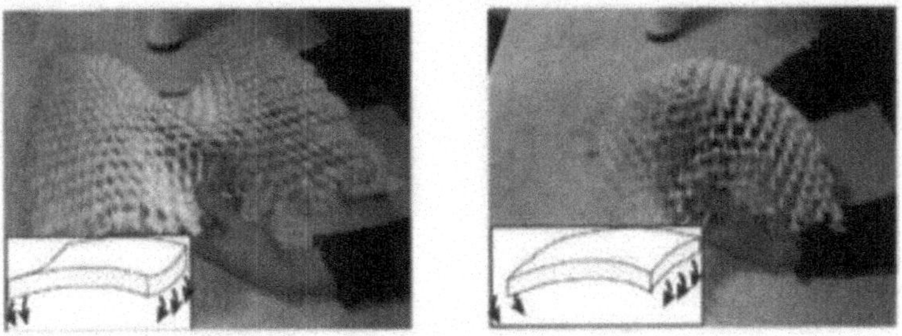

(a)Traditional materials (b)Negative Poisson's ratio material

Fig. 1.23 Bending deformation under the action of out-of-plane bending moment

(7) Band gap frequency

By analyzing the bandgap capability of different honeycomb structures, Shu-Tian Liu and Yu Huang found that the dilated honeycomb structure has a wider bandgap frequency compared with other conventional honeycomb structures.

(8) Deformation behavior

Under the action of out-of-plane bending moment, the deformation characteristic of positive Poisson's ratio material is saddle-shaped surface, as shown in Fig. 1.23a; while the deformation characteristic of negative Poisson's ratio material is arch shape with isotropic curvature, as shown in Fig. 1.23b. The negative Poisson's ratio material shows an arch shape without excess external force, which will effectively avoid damage to the material and has good prospects for application in the wing panel and nose cone structure of aircraft.

(9) Fracture toughness

The fracture behavior of both conventional foam materials and negative Poisson's ratio materials in the tensile state show the mechanical properties of brittle materials, i.e., the materials do not exhibit significant necking, and fracture failure usually occurs suddenly, as shown in Fig. 1.24.

(10) Permeability

The special deformation ability of negative Poisson's ratio materials, which expand outward when stretched and contract inward when compressed, alters their permeability characteristics, showcasing unique properties in filtration applications. Alderson et al. were the first to demonstrate the open-pore characteristics of auxetic materials, which enable the passage of substances from macro to nano scales, thereby

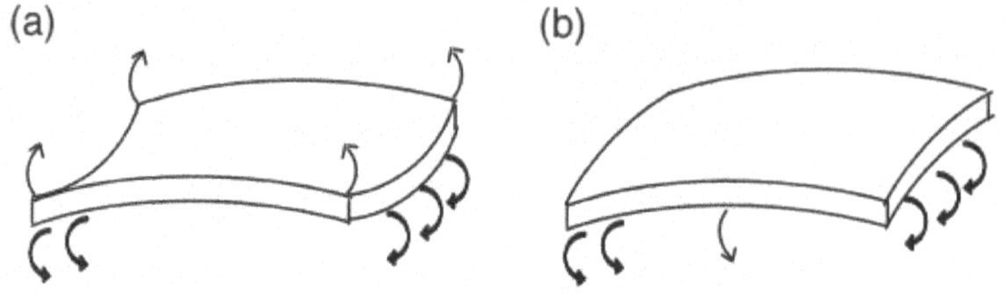

(a)Traditional materials (b)Negative Poisson's ratio material

Fig. 1.24 Bending deformation under the action of out-of-plane bending moment

enhancing their filtration performance. When subjected to increased pressure or uniaxial loads, negative Poisson's ratio polymeric materials are capable of enlarging the dimensions of the permeable pores, allowing for the adjustment of pore shape.

(11) Designability

In 2012, Abdelaal firstly investigated the influence of two parameters closely related to the performance of lattice structures, including relative density and structural form, on the performance of negative Poisson's ratio structures from both theoretical and experimental studies, followed by numerical simulation and processing of gradient negative Poisson's ratio structures, and finally explored the application of negative Poisson's ratio structures in biotechnology. The mechanical response of some negative Poisson's ratio structures is closely related to their geometrical characteristics, so the structural forms that meet certain conditions can be obtained through optimal design in engineering practice, such as explosion-proof curtains, intelligent bandages, etc.

(12) Thermal deformation performance

Lim conducted research on the influence of negative Poisson's ratio values in structures on the distribution of thermal stresses in flat plates subjected to surface-applied gradient temperature loads. The study showed that, with Young's modulus or shear modulus remaining constant, thermal stresses decrease with an increase in negative Poisson's ratio values. Conversely, with bulk modulus remaining constant, thermal stresses increase as negative Poisson's ratio values increase. Therefore, in most cases, negative Poisson's ratio leads to a decrease in thermal stresses. As a result, negative Poisson's ratio structures may be a favorable choice for thermal protection compared to materials with low modulus and low linear expansion coefficients.

Traditional foam materials and auxetic foam materials both exhibit brittle fracture behavior in tension, where the material suddenly fails without significant necking. Choi et al. demonstrated that fracture toughness is a function of volume compression ratio, with stiffness increasing as the volume compression ratio increases. According to elastic theory, as the Poisson's ratio transitions from positive to negative, the bulk modulus relatively decreases, leading to an increase in volume compression ratio.

Therefore, the fracture toughness of negative Poisson's ratio materials also increases. Choi et al. experimentally confirmed that the fracture toughness of auxetic foam can be 2.6 times that of regular foam (Fig. 1.25).

The unique mechanical and physical properties of negative Poisson's ratio materials provide new directions and insights for the development of special-purpose materials and structures. Particularly, due to the advancing technology for synthesizing and fabricating negative Poisson's ratio materials, they have broad prospects in aerospace, defense, and industrial fields. Negative Poisson's ratio materials, because of their high-volume change, can find applications in sensor design. For example, the negative Poisson's ratio piezoelectric ceramic sensor designed by the U.S. Navy (as shown in Fig. 1.26) exhibits significantly improved sensitivity compared to traditional sensors with positive Poisson's ratio matrices. The enhancement in sensitivity can be an order of magnitude higher. In the case of piezoelectric devices, using negative Poisson's ratio materials as electrodes can significantly enhance their sensitivity.

Negative Poisson's ratio materials also have important applications in biomedicine. Negative Poisson's ratio artificial blood vessels can thicken the walls of blood vessels when internal pressure is generated by blood flow, reducing the

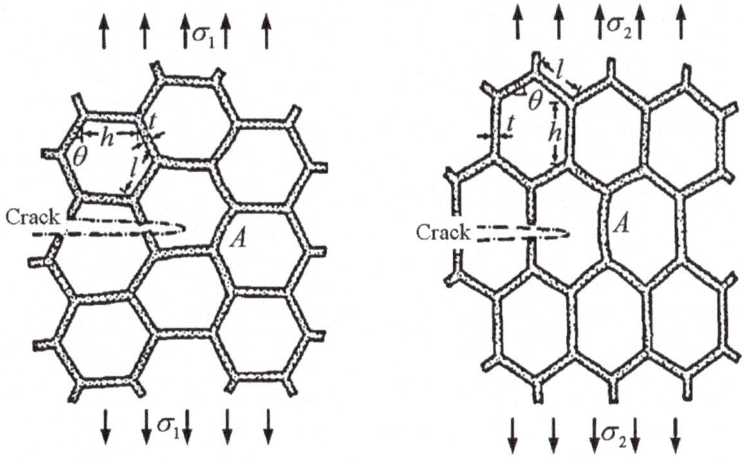

Fig. 1.25 Brittle fracture failure phenomenon of foam material

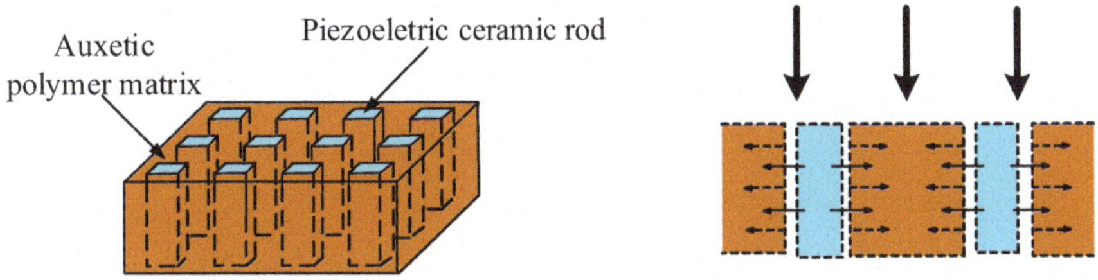

Fig. 1.26 Negative Poisson's ratio piezoelectric ceramic sensor

possibility of vessel rupture. Negative Poisson's ratio materials can also be used in biomedical applications such as arterial dilators, implant surgery and suturing (Fig. 1.27).

Negative Poisson's ratio materials also have significant applications in the aerospace industry and defense. In 1963, Garber et al. applied negative Poisson's ratio pyrolytic graphite in the thermal protection of aerospace vehicles. Similarly, materials with negative Poisson's ratio properties have been used in the blades of turbofan engines.

In the defense sector, the incorporation of negative Poisson's ratio materials into combat suits and protective equipment can reduce weight while providing equivalent protection, thereby improving comfort and agility. Mitsubishi Heavy Industries has designed zero Poisson's ratio bullets, which incorporate negative Poisson's ratio materials, to reduce lateral expansion during bullet firing, thereby enhancing bullet performance (Fig. 1.28).

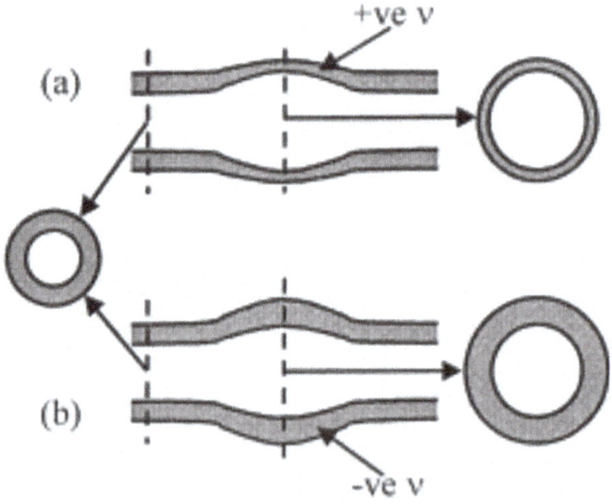

Fig. 1.27 Artificial blood vessel **a** Traditional material **b** Negative Poisson's ratio material

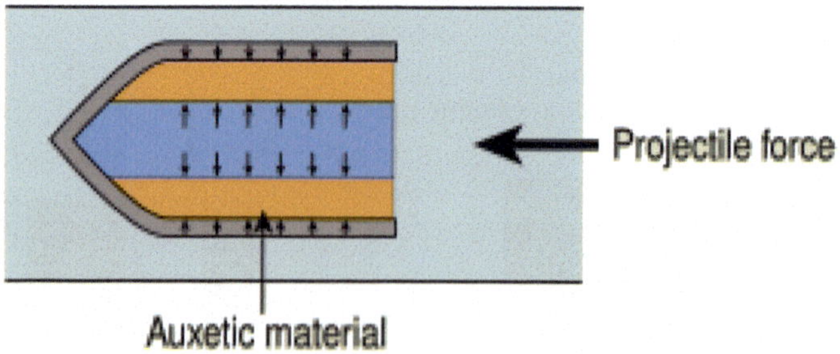

Fig. 1.28 Zero poisson's ratio bullet

1.5 Study on Mechanical Properties of Negative Poisson's Ratio Structure

Based on the micro-unit structure of honeycomb with a porous network connection, negative Poisson's ratio materials and structures exhibit unique negative Poisson's ratio characteristics different from traditional materials when subjected to external stretching or bending. Their special mechanical properties mainly include: (1) When stretched, negative Poisson's ratio materials expand instead of the traditional contraction, and when compressed, they contract perpendicular to the direction of force instead of expanding; (2) The material has better energy absorption characteristics under external impact; (3) When the material is bent, a hollow low-pressure zone is formed inside, which can significantly enhance the back support of the material; (4) While ordinary materials easily crack due to stress concentration, causing material failure, negative Poisson's ratio materials can effectively enhance the strength at the crack and mitigate the phenomenon of cracks in the material; (5) Compared with traditional structures and materials, negative Poisson's ratio structures have a better vibration and noise reduction effect; (6) Negative Poisson's ratio materials have higher shear modulus and resilience.

Classical mechanics reveals that the Poisson's ratio effect will influence the manifestation of other properties of the material. When the structure has a negative Poisson's ratio characteristic, several changes occur in the mechanical performance of the material. For general traditional materials, there is a specific relationship between Young's modulus and shear modulus, with Young's modulus typically being more than double the shear modulus. However, when a material has a negative Poisson's ratio effect, the shear modulus and Young's modulus gradually converge. When the Poisson's ratio is -0.5, they are equal; at this point, the material is easily compressible but difficult to shear. Traditional materials exhibit anti-curvature, i.e., the transverse curvature is opposite to the primary curvature when subjected to bending moments. In contrast, negative Poisson's ratio materials exhibit co-curvature, where the transverse and primary curvatures align. Additionally, while traditional materials contract in the direction of external axial impact and expand perpendicular to that direction (resulting in a positive Poisson's ratio), negative Poisson's ratio materials contract and flow towards the point of impact (resulting in a negative Poisson's ratio). This behavior significantly enhances the material's resistance to indentation. Finally, compared to traditional materials, the expansion of micro-units in negative Poisson's ratio materials and structures under external stretching greatly enhances their crack resistance.

To delve deeper into the unique properties of negative Poisson's ratio structures and materials and expand their range of applications, researchers have extensively studied these materials using mathematical derivations, Monte Carlo methods, finite element analysis, topological optimization, and advanced experimental equipment. Evans conducted experimental comparative studies on the resilience of negative Poisson's ratio foam materials versus traditional foams. The research indicated that the Young's modulus of traditional foam increased only within a specific range during the drop ball test, followed by rapid elastic collapse. In contrast, the resilience of negative Poisson's ratio foam materials showed an increasing trend. Chan conducted tensile, compression, and shear tests on traditional foam materials and negative Poisson's ratio foam materials, further demonstrating that the elastic modulus of negative Poisson's ratio foam materials increases faster under compression compared to traditional foam materials. Scarpa used the finite element method to compare the viscoelasticity of general traditional materials and negative Poisson's ratio materials. The study pointed out that compared to traditional materials, negative Poisson's ratio materials have higher storage moduli and static moduli. Zhang Xinchun et al. used the finite element tools Ansys/Ls-Dyna for an in-depth study of the dynamic impact performance of negative Poisson's ratio honeycomb materials under in-plane impact and explored the effects of impact speed and cell expansion angle on material deformation and energy absorption capability. Qi Chang et al. proposed a negative Poisson's ratio car seatbelt weave due to the unique mechanical properties of negative Poisson's ratio materials, and compared the performance differences between negative Poisson's ratio seatbelts and traditional seatbelts, indicating that negative Poisson's ratio seatbelts can better reduce the chest compression deformation during collisions. Yan Fangfang et al. researched the impact of negative Poisson's ratio microstructure parameters on the material's deformation capacity and bearing capacity. Lu Chao et al. studied the elastic modulus of negative Poisson's ratio honeycomb cores under large deformation conditions, providing a useful reference for the engineering application of negative Poisson's ratio honeycomb materials.

Due to their unique micro-topological structure, different from traditional materials, negative Poisson's ratio structures and materials possess unmatched physical and mechanical properties in many areas, such as significantly enhanced shear modulus, crack resistance, and resilience. These excellent mechanical properties make them promising in sectors like defense, aviation, automotive, and electronics. Researchers have made significant advancements in studying the mechanical properties of negative Poisson's ratio materials over the past few decades. However, current research on negative Poisson's ratio structures and materials still faces issues such as low research efficiency and incomplete theoretical understanding, requiring further efforts by scholars.

1.6 Microstructure of Negative Poisson's Ratio Materials

With an increasing number of scholars beginning to research negative Poisson's ratio structures, a vast amount of natural negative Poisson's ratio materials have been discovered, and artificial ones have been fabricated and synthesized over the past thirty years. Different types and scales of negative Poisson's ratio materials have various application fields and methods. The types of materials primarily include metals, ceramics, composites, and polymers. The scale of the materials ranges from the molecular level, the micro level, to the macro level. Figure 1.29 illustrates the negative Poisson's ratio structures and materials at different scales.

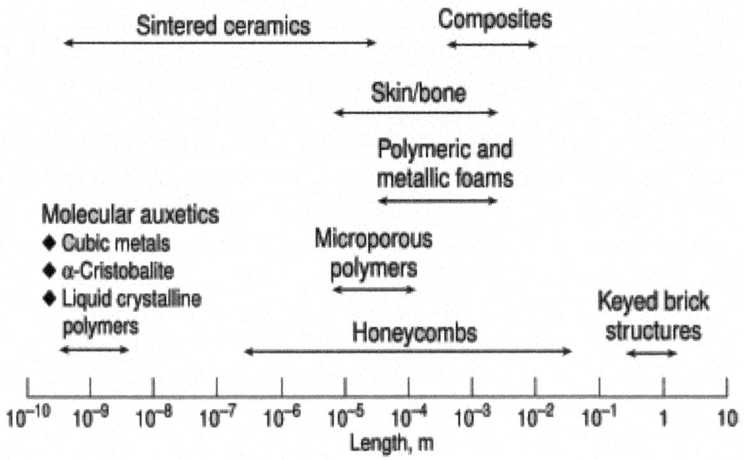

Fig. 1.29 Macroscopic to molecular scale negative Poisson's ratio materials and structures

From the chart below, metals, liquid crystal polymers, and α-cordierite belong to the molecular scale negative Poisson's ratio materials, with a size ranging from 10^{-10} m to 10^{-8} m. Microscale negative Poisson's ratio materials mainly consist of general honeycomb materials, microporous polymers, and composites, typically ranging from 10^{-7} to 10^{-1} m. Macroscale negative Poisson's ratio materials primarily encompass concave structures, wedge mechanisms, and porous honeycomb structures, typically ranging from 10^{-4} to 10^{1} m. The differences in the base materials and scale levels cause variations in the properties of the negative Poisson's ratio materials. Additionally, researchers employ different methods for studying negative Poisson's ratio structures and materials at these various scales.

1.6.1 Microstructural Deformation Mechanism of Negative Poisson's Ratio Materials

The classical elasticity theory fails to explain the inherent special properties of certain materials. It has been found that the microstructure of negative Poisson's ratio materials has a significant impact on their unique mechanical properties. Therefore, when studying the characteristics and deformation mechanisms of negative Poisson's ratio materials, the microstructure of the materials must be taken into consideration. In recent years, scholars have utilized various methods to extensively research the microstructure and deformation mechanisms of negative Poisson's ratio structures to explore the fundamental reasons for their special properties, aiming to better design materials with desired performance.

Lakes, as previously mentioned, manufactured negative Poisson's ratio foam structures through thermo-mechanical treatment methods. Upon studying its microstructure, it was found that the microstructure of negative Poisson's ratio structures greatly differs from traditional foam microstructures. Negative Poisson's ratio foam materials have honeycomb micro units with a concave structure, whereas traditional foams exhibit straight cell wall structures. Evans pointed out that the concave angles between cell walls are the primary reason for the material's unique properties. Negative Poisson's ratio effects can be achieved through changes in the angle of cell walls, variations in wall thickness, wall bending, folding, and the hinge effect. Evans also believed that the folding of cell walls is the main reason for the negative Poisson's ratio characteristics under high strains. Additionally, the buckling of the walls has a significant impact. When the material is subjected to shear forces, the negative Poisson's ratio characteristics of the material are mainly achieved through wall stretching, bending, and hinging. To further study the microstructure and deformation mechanism of negative Poisson's ratio, Evans simulated the cross-sectional micro units of the materials using computers, with unit side lengths of one-billionth of a millimeter. The research found that when the material is subjected to tensile forces, it tilts unidirectionally towards the open end, with some fibers aligning in the direction of external force, thereby increasing the distance between fibers and causing the material to exhibit negative Poisson's ratio characteristics. JeffMore's study suggested that the interconnection at the micro level of negative Poisson's ratio materials is similar to a bow with two arrows. When stretched, the molecular chains align laterally, displaying negative Poisson's ratio characteristics. Alderson et al. elucidated the mechanism of the negative Poisson's ratio effect by constructing a node-raw filament model. They believed that, under low strain, the hinge produced between the raw filaments when external force is applied is the reason for the material's negative Poisson's ratio characteristics. Griffin et al. believed that, during molecular-scale research on negative Poisson's ratio structures, the distance between polymer main chains increases when subjected to external tensile forces, causing rod-shaped molecular structures to be perpendicular to the direction of external forces. When rod-shaped molecules are long enough and the van der Waals forces between them can't prevent them from re-arranging, the material expands in a direction perpendicular to the external force, displaying negative Poisson's ratio characteristics. Milton and colleagues believed that negative Poisson's ratio materials have a multi-level structure. The microstructures are hierarchically arranged, and each level contains microstructures that are also hierarchically ordered. The size of the hierarchical components and the order of the microstructure arrangement are fundamental reasons for the unique mechanical properties of negative Poisson's ratio materials. Additionally, to explain the mechanisms of the negative Poisson's ratio characteristics, some researchers have tried to apply the continuity mechanism and Cosserat mechanism, believing that the continuity mechanism is the fundamental reason for the special mechanical properties of negative Poisson's ratio materials. Some also believe that the unique mechanical properties of negative Poisson's ratio materials are due to the lateral deformation caused by the material when subjected to axial external forces.

Researchers have extensively studied the microstructures and deformation mechanisms of negative Poisson's ratio materials of different materials and scales. This has deepened our understanding of the characteristics and origins of negative Poisson's ratio materials, laying a foundation for subsequent researchers to design negative Poisson's ratio material microstructures to meet related performance requirements.

1.6.2 Microstructure Design of Negative Poisson's Ratio Materials

With the advancement of science and technology, there are an increasing number of structural optimization design methods applied in engineering designs, and their efficiency is also improving. Typically, these methods modify the shape of the structure to ensure performance meets specific requirements or achieves an optimal state in a certain sense. While macro design of structures has matured, there are fewer effective methods for optimizing the microstructure of materials. For specific materials like honeycomb structures, the microstructure plays a pivotal role in their performance. Hence, designing a rational microstructure to fulfill specific performance requirements is of paramount importance. Through designing the microstructure of materials, we can obtain composite with specific unique properties, such as optimal damping, negative thermal expansion coefficient, negative Poisson's ratio, and excellent piezoelectric characteristics.

Due to their lightweight characteristics and some specific physical functionalities, honeycomb structures have found extensive applications in defense, transportation, chemical engineering, and electronics. The negative Poisson's ratio structure also falls under the broader category of honeycomb structures. The performance of a honeycomb structure depends on its base material and the microstructure of its unit cell. Establishing a relationship between the macroscopic properties of the material and the parameters of its microstructural unit, and independently designing the microstructure based on performance requirements, is crucial. Research indicates that the geometric parameters of the honeycomb structure's micro-unit, such as edge length and wall thickness, significantly influence the mechanical properties of the honeycomb structure. Furthermore, spatial topological parameters like the number and arrangement of adjacent cells also have an impact on the structure's performance.

Currently, the optimization design of microstructures in composite both domestically and internationally is generally based on mesoscopic mechanics finite element methods. This method was proposed in the 1970s and saw significant development in the late 1980s with the rise of computational materials science. In the 1990s, researchers from both China and abroad conducted extensive studies on this, and it remains a research hotspot in the fields of mechanics and materials science. In 1994, Sigmund employed the homogenization method to optimize the topology of the micro-units of composite. Since the form of the microstructure of the material was unknown, the inverse homogenization theory was applied to determine the best layout of the structure under specific constraints, ensuring the performance was optimized. Song Yingdong and Sun Jie effectively combined the micro-topological structure and macro-mechanical properties of composite using a high-precision general single-cell model. Using a composite material microstructure topology optimization design method based on the genetic algorithm, they optimized the microstructure of the material, achieving superior shear modulus and Poisson's ratio. These optimization methods have greatly facilitated the design of material microstructures. However, challenges such as difficulties in establishing optimization models and low optimization efficiency persist. Thus, seeking more effective methods to optimize the microstructural parameters of negative Poisson's ratio materials to further enhance their performance and promote their application in industrial design and manufacturing remains an important research topic in this field.

1.7 Other New Negative Poisson's Ratio Structures and Their Applications

Xu Manqiong et al. proposed a specimen with negative Poisson's ratio properties. This specimen involves embedding multiple arc-shaped bodies within a base material. The arc-shaped bodies on the left and right sides of the specimen are distributed in a mirror-symmetric manner, with the arc-shaped bodies on the same side having the same curvature and being evenly spaced as shown in Fig. 1.30. The elastic modulus and stiffness of the arc-shaped bodies differ from the base material. During the solidification process of the specimen's base material, arc-shaped bodies can be embedded in the direction of the force exerted on the specimen. Alternatively, after the solidification of the base material, arc-shaped bodies embedded in the force direction can be removed. In another method, arc-shaped bodies can be produced on the entire base material specimen through reinforcement or mechanical processing or can be formed via chemical reactions. The arc-shaped bodies can include fiber-reinforced arc-shaped bodies, bubble-weakened arc-shaped bodies, arc-shaped gaps, multi-level curved arc-shaped bodies, multi-radius curved arc-shaped bodies, and fish scale-like arc-shaped bodies. When the arc-shaped bodies are subjected to tension in their tangential direction, they exhibit expansion in the perpendicular direction. When subjected to compression, they show contraction and squeezing in the perpendicular direction. This specimen can be widely applied in fields such as composite plates, deep-sea operations, aerospace, and aviation.

Pan Zelin and others, based on the microporous structure of natural foam and the principle of generating a negative Poisson's ratio, designed and prepared a biomimetic material with a special negative Poisson's ratio structure. The biomimetic negative Poisson's ratio material of this invention is a porous foam structure material obtained from thermoplastic plastic through a screw extrusion machine. The pores in this material are elongated hive micropores, and numerous elongated hive micropores are arranged radially, layer by layer, diverging outward from the central axis as shown in Fig. 1.31. When this material encounters an impact, the bidirectional contraction

Fig. 1.30 Structure of the specimen

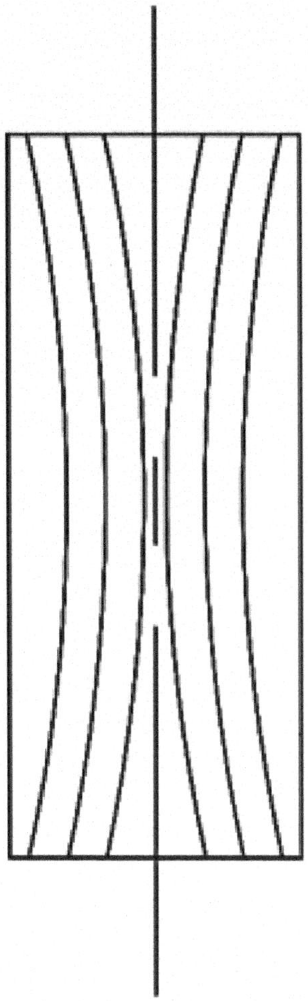

caused by its negative Poisson's ratio quickly converts the impact force into deformation energy, which is absorbed. Simultaneously, the density of the material increases rapidly, thereby enhancing its impact resistance.

Hu Hong proposed a three-dimensional negative Poisson's ratio spacer knitted fabric and its manufacturing method. The three-dimensional negative Poisson's ratio spacer knitted fabric includes a first surface fabric, a second surface fabric, and a spacer yarn layer connecting the first and second surface fabrics. Both the first and second surface fabrics have a negative Poisson's ratio effect as shown in Fig. 1.32. The manufacturing method includes: Using a three-dimensional mesh spacer knitted fabric as the base fabric, Compressing or stretching the base fabric in a planar direction to form a surface structure with a negative Poisson's ratio effect, and using a heat setting method to maintain the negative Poisson's ratio effect of the surface structure of the base fabric formed in the previous step, resulting in the three-dimensional

Fig. 1.31 Structure of the bionic negative Poisson's ratio material

negative Poisson's ratio spacer knitted fabric. The production method of this invention is cost-effective and easy to operate. The negative Poisson's ratio effect of the three-dimensional spacer knitted fabric proposed by this invention is significant, and its resilience is excellent, making it of high practical application value.

Ma Yanhong et al. proposed an inner hollow metal-rubber isolator with negative Poisson's ratio characteristics. This isolator is composed of multiple structurally identical complete unit bodies and several structurally identical supplementary unit bodies arranged in an overlapping manner, as shown in Fig. 1.33. The supplementary unit body is half the structure of the complete unit body. On the one hand,

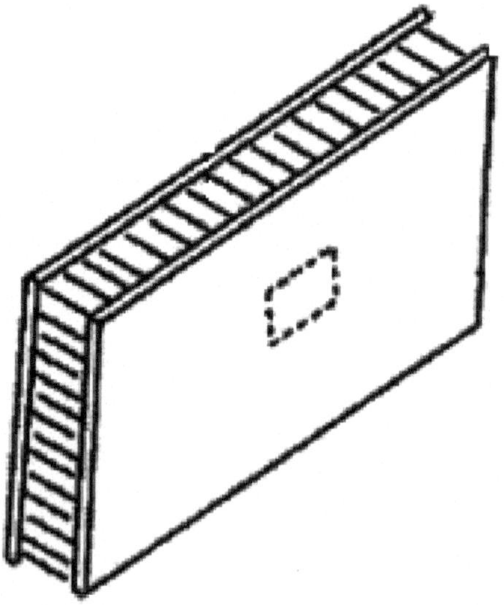

Fig. 1.32 Structure of the three-dimensional negative Poisson's ratio spacer knitted fabric

this invention utilizes the excellent damping performance of metal-rubber structured damping materials to address the issues of vibration fatigue and impact failure in the isolator. On the other hand, by the alternating arrangement of complete unit bodies and complementary unit bodies, the staggered structure brings about the negative Poisson's ratio characteristics, thus enhancing the isolator's capacity to resist impact loads. The invention combines the high-damping structural material of metal-rubber with the features of a negative Poisson's ratio structure. Through the staggered arrangement of multiple unit bodies, an inner hollow metal-rubber isolator with negative Poisson's ratio characteristics is formed. The isolator not only leverages the outstanding damping properties of the metal-rubber structure to address the vibration fatigue and impact failure problems of the inner hollow metal-rubber isolator but also utilizes the negative Poisson's ratio characteristics brought by the staggered structural design to enhance its ability to withstand impact loads.

Yao Yongtao et al. introduced an expandable reflector mirror body based on the negative Poisson's ratio of a chiral honeycomb structure with double curvature, as shown in Fig. 1.34. This invention consists of a laying layer and a chiral honeycomb structure layer with double curvature and a negative Poisson's ratio. The shape of the structure layer is parabolic, and its inner surface serves as the laying layer. The material of the structure layer is made from shape memory polymer materials, while the laying layer is composed of metal materials as well as silicon carbide or silicon carbide-based composites. The purpose of this invention is to address the issues of current aerospace reflectors, which cannot achieve a large expansion volume to

Fig. 1.33 Structure of the inner hollow metal-rubber vibration isolator

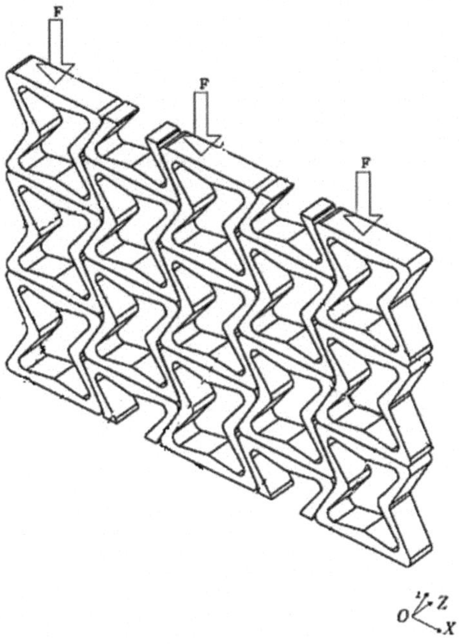

contraction volume ratio, are not lightweight enough, have a small aperture, and possess complex control systems. This invention boasts a high expansion volume to contraction volume ratio and the capability of multiple repeated foldings and expansions. It can be unfolded and formed at once, and its structure is simple. Moreover, it exhibits a high degree of lightness, and the system is highly reliable and stable.

Qi Chang et al. have disclosed a novel car safety belt webbing designed to prevent chafing injuries. It is composed of M × N expansion units, presenting as a continuous belt-like structure with orthogonal protrusions, as shown in Fig. 1.35. Here, M represents the number of columns of expansion units contained in the webbing, and N stands for the number of expansion units in a complete column of the webbing. The shape of the expansion unit is a continuous curved surface with orthogonal protrusions, comprising four orthogonal protrusions and four nearly triangular transition surfaces. When a vehicle collision occurs, the safety belt is hit by the human torso and upper limbs, which have acceleration. The webbing stretches longitudinally. Due to its negative Poisson's ratio expansion properties, the webbing expands laterally, increasing the contact area between the webbing and the human body. Under the same impact force, the enlarged contact area reduces the pressure exerted on the chest and abdomen of the occupant, thus reducing the compression on the passenger's torso. This helps avoid injuries to the ribs and internal organs, preventing secondary injuries to the passengers.

Ma Xiqin et al. have disclosed a damping plate based on a negative Poisson's ratio structure. It consists of a base plate and damping material. A negative Poisson's ratio structure, which forms several cavities in its cross-section, is affixed to the base plate. Cavities are also formed between the negative Poisson's ratio structure and the base plate. The damping material, used as filler, is filled into these cavities, as shown in Fig. 1.36. This invention leverages the tensile and compressive deformation characteristics of the negative Poisson's ratio structure. By allowing the material farthest from the base plate to deform fully, the damping material is effectively utilized, and the dissipation effect of the vibrational energy is significantly increased. This

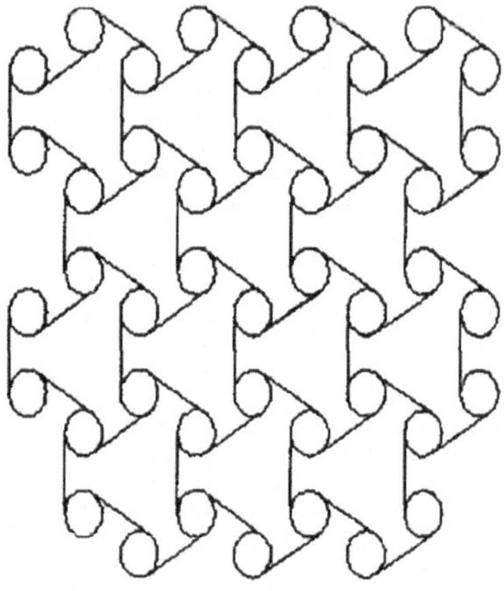

Fig. 1.34 Structure of deployable reflector mirror body

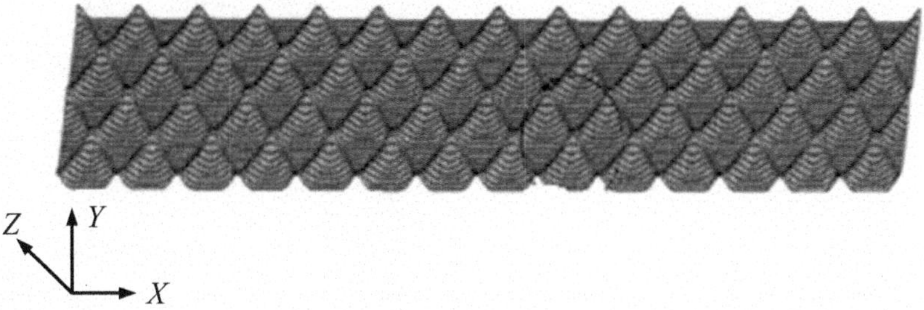

Fig. 1.35 Structure of the anti-strangulation vehicle seat belt webbing

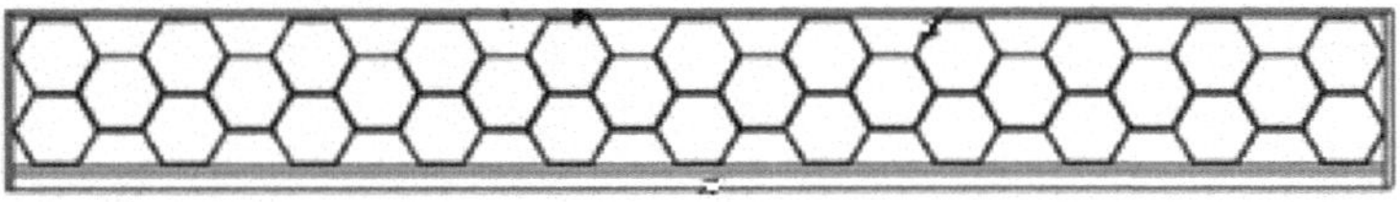

Fig. 1.36 Structure of the damping plate

addresses the limitations of conventional damping plates, notably optimizing the vibrational characteristics of the damping plate. In light of the aforementioned shortcomings in existing technology, the aim of this invention is to provide a damping plate based on a negative Poisson's ratio structure that ensures full material deformation and notable vibrational energy dissipation effects. The damping plate based on a negative Poisson's ratio structure includes a base plate and damping material, with the negative Poisson's ratio structure affixed to the base plate forming several cavities in its cross-section and between itself and the base plate, with the damping material filling these cavities.

Huang Xiangdong et al. proposed a buffer block structure for shock absorber assembly as shown in Fig. 1.37. It comprises a top, a base, and a negative Poisson's ratio component positioned between the top and the base. The negative Poisson's ratio component consists of multiple layers of negative Poisson's ratio modules. Each negative Poisson's ratio module includes a supporting layer and a connecting layer attached to the supporting layer. The supporting and connecting layers of adjacent negative Poisson's ratio modules are interconnected. The negative Poisson's ratio module contains multiple negative Poisson's ratio units, where each unit is a hollow structure formed by the connecting layer enclosing the supporting layer. In addition, the invention also discloses a shock absorber assembly. By employing the shock absorber assembly and buffer block structure of this invention, the design is rational, achieving the desired non-linear stiffness values. This enhances the smoothness of vehicle movement and further improves the buffer block's ability to resist impact loads.

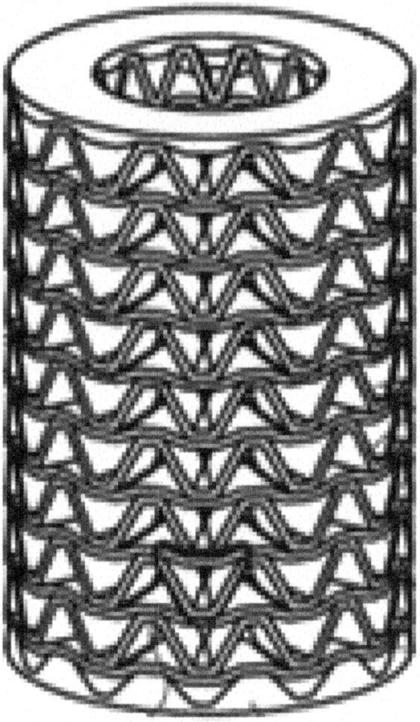

Fig. 1.37 Structure of the cushion block

Liu Wangyu et al. introduced a buffer energy-absorbing structure as depicted in Fig. 1.38. It comprises an outer shell. The buffer energy-absorbing structure filled within the shell consists of multiple layers of metal grid structures or porous foam structures. Adjacent layers of the metal grid or porous foam structures alternately stack in a longitudinal direction with positive and negative Poisson's ratios. The metal grid structure with a positive Poisson's ratio is composed of a three-dimensional honeycomb grid made up of an array of regular hexagonal cells. In contrast, the metal grid with a negative Poisson's ratio consists of a three-dimensional stretching grid formed by an array of concave cells. The porous foam structure with a positive Poisson's ratio has holes composed of an array of diamond structures, whereas the porous foam structure with a negative Poisson's ratio features holes in the form of an array of four-pointed star structures. Compared to conventional buffer energy-absorbing structures, this design can effectively reduce the peak of impact forces, making the entire energy-absorption process smoother and thereby enhancing the energy-absorption efficiency of the structure. Collision safety is a critical metric for the safety of various vehicles, with a significant portion of materials in the vehicle's buffer collision zone dedicated to resisting impacts. Efficient impact energy absorption, reduced weight, and minimized peak loads are the goals pursued in the structural design of these functional parts. The purpose of this invention is to overcome the shortcomings and limitations of existing technologies. It aims to provide a buffer energy-absorbing structure that improves the deformation pattern of existing energy-absorbing structures and enhances their efficiency. When compared to traditional buffer energy-absorbing structures, it can effectively reduce the impact force's peak,

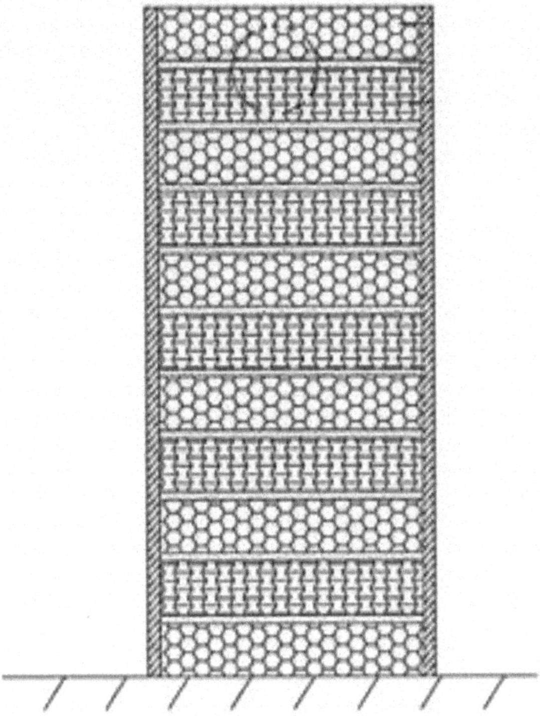

Fig. 1.38 A buffered energy-absorbing structure using the negative Poisson's ratio materials

leading to a smoother energy-absorption process and thus heightening the structure's energy-absorbing efficiency.

Ma Pibo et al. introduced a negative Poisson's ratio two-dimensional warp-knitted fabric and its weaving method as depicted in Fig. 1.39. The negative Poisson's ratio two-dimensional warp-knitted fabric consists of two parts: the ground tissue part and the deformation part. The weaving method involves using a warp-knitted hexagonal mesh fabric as the base fabric. An additional comb is used during weaving to exert force on the base fabric in the plane direction, causing the base fabric structure to become unbalanced and skew to the left and right. This results in a warp structure with a negative Poisson's ratio effect, producing the negative Poisson's ratio two-dimensional warp-knitted fabric. The weaving method proposed in this invention is cost-effective and straightforward to operate. The two-dimensional warp-knitted fabric of this invention exhibits a pronounced negative Poisson's ratio effect, and its weaving method is simple. The practical application range of the two-dimensional fabric is broader and can also be extended to the three-dimensional fabric domain. By altering the warp delivery amount during weaving, the hexagonal mesh ground tissue of the fabric undergoes a serrated left–right skew deformation, thereby creating a negative Poisson's ratio effect.

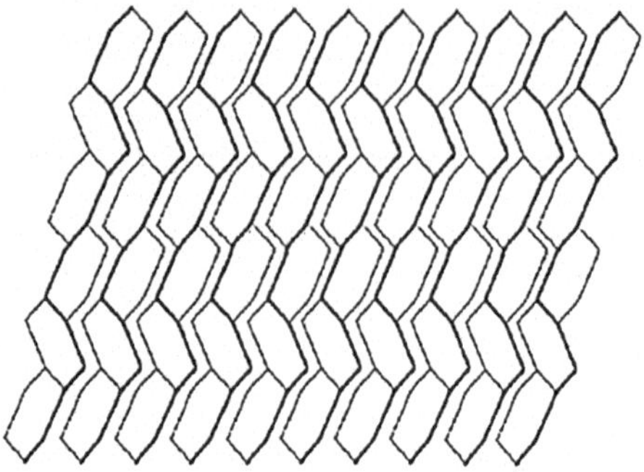

Fig. 1.39 Structure of the negative Poisson's ratio two-dimensional warp knitted fabric

Tan Huifeng et al. proposed a negative Poisson's ratio honeycomb structure for an expandable shell based on bistable composite, as shown in Fig. 1.40. The structure is composed of several honeycomb cell structures arranged in a staggered manner. The honeycomb cell structure consists of a bistable composite material shell honeycomb cell structure formed by bistable shell rods and connecting tubes. One end of the four bistable composite material shell rods is connected to a connecting tube with a curling stable curvature radius consistent with the shell rod, while the other end is connected to the connecting tube of another group of cells, forming a honeycomb structure with negative Poisson's ratio characteristics. Unlike the honeycomb cell structures supported by flat plates, this invention uses composite material shell rods with bistable characteristics. This enhances the support stiffness of the formed honeycomb structure. During compression, it automatically curls and shrinks, avoiding the extensive buckling phenomena observed with flat structures. This greatly enhances the negative Poisson's ratio effect and structural rigidity of such structures, improving their reliability. While negative Poisson's ratio structures have a wide range of applications, their study remains largely theoretical. The challenge is to enhance its rigidity while maintaining its negative Poisson's ratio effect. 3D printing technology has boosted the application development of these structures, but rigidity improvement is still a weak link. By adopting bistable composite material expandable shell rods, it's expected to enhance both the negative Poisson's ratio effect and rigidity of these structures, promoting their application. The objective of this invention is to introduce a bistable composite material expandable shell into a negative Poisson's ratio honeycomb structure. It provides a honeycomb structure based on a bistable composite material expandable shell with high rigidity, stable curling state, and pronounced negative Poisson's ratio effect, offering support for the application of such structures.

Yin Guangsheng et al. introduced an expansion joint device with a negative Poisson's ratio honeycomb structure, as shown in Fig. 1.41. The device includes several bridge beams. Above the beams, there is a pavement layer. A gap is left between the ends of adjacent beams, and within this gap, a honeycomb structure with a negative Poisson's ratio effect is installed and connected to the beam ends. Above the honeycomb structure, there is an asphalt paste that connects with the pavement layer, and the upper surface of the asphalt paste is flush with the upper

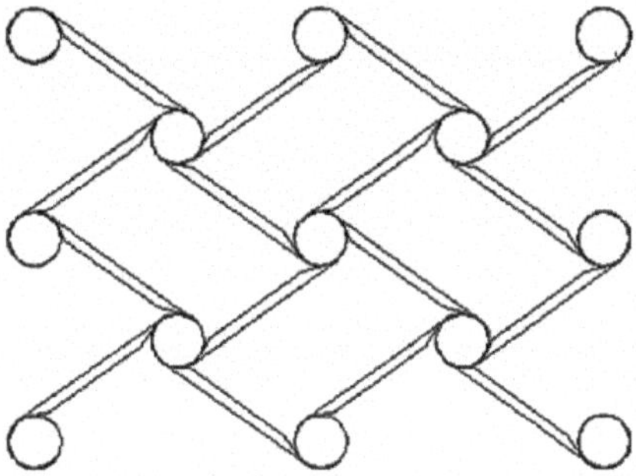

Fig. 1.40 The honeycomb cell structure

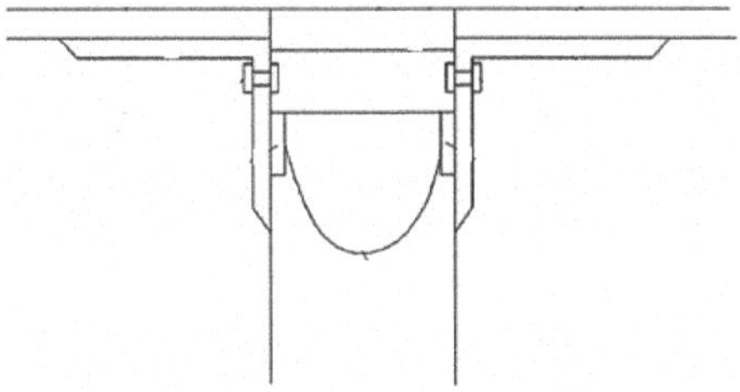

Fig. 1.41 Structure of the expansion joint device

surface of the pavement layer. By installing the honeycomb structure with a negative Poisson's ratio effect, the structure contracts when compressed and expands when pulled. Within the gap between beams, the honeycomb structure also possesses a certain load-bearing capacity both horizontally and vertically. This design not only allows for the free longitudinal deformation between adjacent beam ends but also provides adequate restraint to prevent relative misalignment of the beams. With the application of the asphalt paste, the smooth passage of vehicles is ensured, and it also serves a drainage protection function. This new practical design features a simple structure, and it boasts excellent durability and elasticity.

Liu Kaiye proposed a honeycomb sandwich structure with a negative Poisson's ratio, suitable for applications of honeycomb sandwich structures under conditions of high curvature, as shown in Fig. 1.42. The innovative design of the honeycomb sandwich structure reshapes the honeycomb into a deformed "T" shape. The middle section is recessed, with the upper part being a recessed inverted trapezoid, while the lower indentation takes a triangular form. When the laminate is bent, the two ends of the "T" shape bend outward, while the central part contracts inward under force, thereby exhibiting a negative Poisson's ratio. This design is particularly suited for surfaces with high curvature. In engineering, many structures require the creation of honeycomb sandwich structures with a significant curvature. Conventional hexagonal honeycomb laminates under high curvature conditions can struggle to resist large in-plane deformations during fabrication and use, leading to saddle point effects. However, honeycomb structures with a negative Poisson's ratio are less likely to encounter such issues when adapting to complex curved surfaces. To overcome the

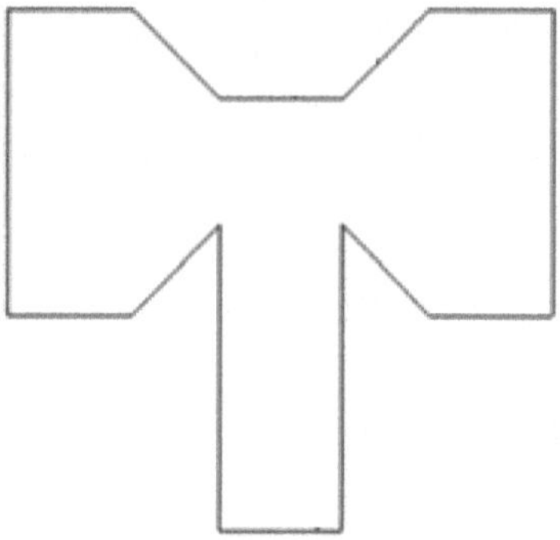

Fig. 1.42 Structure of the honeycomb interlayer

limitations of existing technology, this application introduces a new honeycomb sandwich structure. This design, featuring the deformed "T" shape with its middle section recessed, an upper inverted trapezoid, and a lower triangular indentation, showcases a negative Poisson's ratio when bent, making it well-suited for surfaces with significant curvature.

Liu Yanju et al. proposed a deformable skin structure with a designable negative Poisson's ratio, as illustrated in Fig. 1.43. This structure pertains to an aircraft wing skin design. It aims to address the challenge where existing skin structures cannot meet the requirement of having a low deformation stiffness in one specific direction while maintaining a high structural stiffness in other directions to bear aerodynamic loads and inertial loads. The structure includes a honeycomb structure matrix, a flexible elastomer, and multiple pneumatic artificial muscles. The honeycomb structure matrix is composed of a mix of positive Poisson's ratio honeycomb units and negative Poisson's ratio honeycomb units. The flexible elastomer is poured onto the honeycomb structure matrix and multiple pneumatic artificial muscles, forming a flexible panel. This invention is used for fabricating aircraft wing skins.

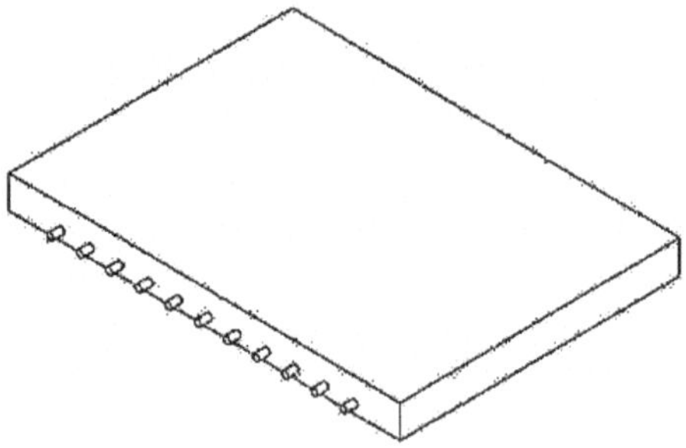

Fig. 1.43 The deformation skin structure with a designable negative Poisson's ratio

1.8 Summary of This Chapter

In this chapter, the important research progress of the development process of negative Poisson's ratio structure is introduced firstly. Then this chapter classifies the negative Poisson's ratio structure through materials, focusing on the research progress of porous negative Poisson's ratio materials, negative Poisson's ratio polymerization and composite materials. In terms of theoretical research, this chapter briefly introduces the mechanical properties of negative Poisson's ratio and its mechanical properties research methods, and related research on the microstructure of negative Poisson 's ratio, paving the way for subsequent chapters. In terms of practical applications, this chapter summarizes the new negative Poisson's ratio materials, and introduces their applications in different engineering designs in different fields, showing the broad application prospects of negative Poisson's ratio materials.

References

1. Ma Yanxuan, Xu Yaxi, Yu Xia et al. Negative Poisson's ratio design and static load mechanical properties of foam concrete [J]. Material introduction 2021, 35(24): 24068–24074.
2. Liu Kai, Cao Xiaofei, Li Ying, et al. Research progress of deformable wing based on chiral superstructure design [J]. Aerospace science and technology. 2022, 33(01): 20–36.
3. Lakes R. Foam structures with a negative Poisson's ratio[J]. Science, 1987, 235(4792):1038–1040
4. Caddock B D, Evans K E. Microporous materials with negative Poisson's ratios. I. Microstructure and mechanical properties[J]. Journal of Physics D : Applied Physics, 1989, 22(12) : 1877–1882.
5. Evans K E, Caddock B D. Microporous materials with negative Poisson's ratios. II. Mechanisms and interpretation[J]. Journal of Physics D : Applied Physics, 1989, 22(12) : 1883–1887.
6. P. Verma, C. He, A. C. Griffin, Phys. Status Solidi B 2020, 257, 2000261.
7. Yang Wubo, Yang Xia, Li Zhongming and so on. Structure and Properties of Negative Poisson's Ratio Materials [J]. Polymer Materials Science and Engineering. 2001(06): 15–18+24. https://doi.org/10.16865/j.cnki.1000-7555.2001.06.004.
8. Gibson LJ, Ashby MF, Schajer GS, et al. The mechanics of twodimensional cellular materials. Proceedings of the Royal Society of London. A. Mathematical and Physical Sciences, 1982, 382(1782).
9. L. Rothenburg, A. A. Berlin and R. J. Bathurst, Microstructure of isotropic materials with negative Poisson's ratio, Nature, 1991, 354, 470–472.
10. MA Zhengdong. Three-dimensional auxetic structures and applications there of : US, US7910193B2[P]. 2011–03–22.
11. Shi Wei, Yang Wei, Li Zhongming, et al. Research progress of materials with negative Poisson's ratio [J]. 2003 (6): 48–57.
12. Herakovich C T.J Compos Mater 1985 18:447.
13. Milton G J.Mech Phys Solids 1992 40:1105.
14. Theocaris P S Stavroulakis G E Panagiotopoulos P D. Archive of Appl Mech 1997 67 : 274.
15. Wei G Y Edwards S F. Physica A 1998 258 : 5.
16. Stagni L.Composite Science and Technology 2001 61:1729.
17. Evans K E Nkansah M A Hutchinson I J et al. Nature 1991 353 : 124.
18. Bowick M Cacciuto A Thorleifsson G et al.Phys Rev Lett 2001 87(14):148103–1.
19. Baughman R H Stafström S Cui C X et al. Science 1998 279 : 1522.

20. Rovati M.Scripta Materialia 2003 48:235.
21. Bezazi A, Boukharouba W, Scarpa F. Mechanical properties of auxetic carbon/epoxy composites: Static and cyclic fatigue behaviour. Physica Status Solidi (B), 2009, 246(9): 2102–2110.
22. Bianchi M, Scarpa F, Smith CW. Shape memory behaviour in auxetic foams: Mechanical properties. Acta Materialia, 2010, 58(3): 858–865.
23. Gibson LJ, Ashby MF, Schajer GS, et al. The mechanics of two-dimensional cellular materials. Proceedings of the Royal Society of London. A. Mathematical and Physical Sciences, 1982, 382(1782): 25.
24. Lakes, R. S., & Witt, R. (2002). Making and characterizing negative Poisson's ratio materials. International Journal of Mechanical Engineering Education. 30(1), 50–58. https://doi.org/10.7227/IJMEE.30.1.5.

Chapter 2
Performance Analysis of Negative Poisson's Ratio Structures

2.1 Introduction

Based on the analysis of the negative Poisson's ratio structures in the previous chapter, it is known that the special mechanical properties of the negative Poisson's ratio structures and materials mainly originate from the structural deformation of the micro units. Therefore, unlike the mechanical properties of conventional materials that depend solely on the properties of the basis material, the deformation mode of the micro units and the basis material in the negative Poisson's ratio structures and materials have important influences on the macroscopic mechanical properties of the materials. Thus, when describing the macroscopic mechanical properties of the negative Poisson's ratio structures, they can be jointly described by the geometric parameters of the micro units and the mechanical properties of the basis material. In the study of negative Poisson's ratio structures and materials, they are usually regarded as homogeneous continuous materials, and their macroscopic mechanical properties are reflected by macroscopic equivalent mechanical models and elastic parameters. By establishing the relationship between microstructure parameters and macroscopic mechanical properties, the mechanical properties of negative Poisson's ratio structures and materials can be better quantitatively described by equivalent elastic parameters such as equivalent elastic modulus and equivalent Poisson's ratio. In expressing the properties of negative Poisson's ratio structures and materials, the relative density is an important parameter used to describe the characteristics of microscopic units, which has an important influence on the equivalent properties of negative Poisson's ratio materials. When describing the structure and properties of negative Poisson's ratio materials, relative density is an important parameter used to describe the characteristics of micro units, which has an important influence on the equivalent performance of negative Poisson's ratio materials. The relative density of negative Poisson's ratio structures is not related to the size of micro units, but

G. Zhou et al., *Design, Manufacturing, and Application of Negative Poisson's Ratio Structure*, https://doi.org/10.1007/978-981-96-1937-5_2

mainly depends on the structure of micro units. However, as a macroscopic statistical quantity, relative density cannot reflect the influence of microstructure. Therefore, studying the relationship between microstructure parameters and macroscopic mechanical properties can better understand the special mechanical properties of negative Poisson's ratio structures. In addition to the static mechanical properties of materials, the relationship between the dynamic mechanical properties of materials and microstructure parameters is more complex, especially the energy absorption process of materials. Negative Poisson's ratio materials have excellent damping and energy absorption effects, and their energy absorption process is mainly achieved by the deformation of microstructure. Therefore, studying the deformation process and energy absorption mechanism of negative Poisson's ratio structures is of great significance for promoting their wide application in industry.

2.2 Double-Arrow Type Negative Poisson's Ratio Structural Mechanics Model

In engineering practice, most materials experience contraction of the cross-section when subjected to uniaxial tension, and expansion of the cross-section when subjected to uniaxial compression. According to the definition of Poisson's ratio, the Poisson's ratio of a material is the ratio of transverse strain to longitudinal strain when subjected to uniaxial compression (or tension). That is

$$\nu = -\frac{\varepsilon_y}{\varepsilon_x} \tag{2.1}$$

Therefore, the definition of Poisson's ratio shows that Poisson's ratio of a material determines the direction and degree of change in the transverse section when the material is subjected to an external tensile or compressive force. Generally, the transverse strain is positive (negative) and the longitudinal strain is negative (positive) when the material is compressed (stretched), which means that the Poisson's ratio of the material is positive, and the Poisson's ratio of most materials in nature is between 0 and 0.5. Contrary to the general material, the negative Poisson's ratio material shrinks in the transverse section when subjected to uniaxial compression, so the transverse strain and longitudinal strain are both negative, and the definition of Poisson's ratio shows that the Poisson's ratio of the material is negative at this time [1]. Figure 2.1 shows the deformation diagram of general material and negative Poisson's ratio material.

The special deformation mode of negative Poisson 's ratio materials makes them different from traditional materials in terms of mechanical properties. In isotropic materials, there are the following relationships between different mechanical properties of materials:

$$E = 2G(1 + \nu) = 3K(1 - 2\nu) \tag{2.2}$$

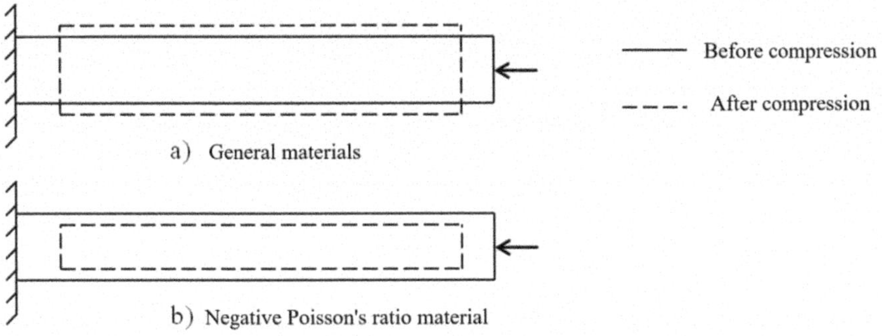

Fig. 2.1 Schematic diagram of deformation in uniaxial compression for general materials and materials with negative Poisson's ratio

$$G = \frac{E}{2(1+\nu)} = \frac{3K(1-2\nu)}{2(1+\nu)} \tag{2.3}$$

E is the Young's modulus of the material, K is the bulk modulus, ν is the Poisson's ratio of the material, and G is the shear modulus. From the formula (2.2), it can be seen that for general materials due to Poisson's ratio is positive, so Young's modulus is generally greater than twice the shear modulus [2]. But when the material Poisson's ratio is negative, the relationship between the two will change, when the material Poisson's ratio is -0.5, Young's modulus and shear modulus equal, the material will be more difficult to shear deformation and easier to compressive deformation. In general, when the Poisson's ratio is close to 0.5, the material's shear modulus will be much smaller than the bulk modulus, the material is easy to shear deformation and difficult to produce compression deformation. For negative Poisson's ratio materials, when the Poisson's ratio is close to -1, from the formula (2.3), the shear modulus of the material will be much larger than the bulk modulus, so the material will be very difficult to shear deformation.

Unlike conventional solid materials, the mechanical properties of multi-cell materials such as negative Poisson's ratio materials are influenced by both the mechanical properties of the base material and the shape of the multi-cell structure. When studying the mechanical properties of such multi-cell structures or materials, they are generally considered as continuous materials and are described by applying the same methods used to describe the mechanical properties of conventional solid materials, called equivalent mechanical properties, such as equivalent Poisson's ratio and equivalent elastic modulus. According to Gibson's research, the mechanical properties of the negative Poisson's ratio structure and the material can be described by the geometric parameters of the micro-unit structure and the mechanical properties of the basis material. The cell unit is the smallest repeated unit of the negative Poisson's ratio structure. When the cell unit of the negative Poisson's ratio structure is subjected to uniaxial compression, the oblique cell wall of the cell structure will deform, and the vertical movement of the oblique cell wall will make the material exhibit negative

Poisson's ratio characteristics. The mechanical properties of the cell will change with the angle of the oblique cell wall. The structure has negative Poisson's ratio characteristics within a certain angle range. The structure has different Poisson's ratio values under different microstructure parameters, and even presents different characteristics with positive and negative Poisson's ratios. By establishing the relationship between the negative Poisson's ratio structure and the geometric parameters of the micro-unit structure and the mechanical properties of the basis material, the parameters can be changed to obtain the change trend of the structural properties, which provides a basis for further understanding the performance of the structure. Therefore, it is of great significance to master the relationship between the macroscopic properties of the structure and the microstructure parameters for the more effective application of the negative Poisson's ratio structure.

2.2.1 Negative Poisson's Ratio Structure Equivalent Elastic Modulus

2.2.1.1 Double-Arrow Type Negative Poisson's Ratio Structure

In their research on designing microstructures, Ulrik and Ole obtained a new negative Poisson's ratio material structure with a Poisson's ratio value of −0.8 by applying a topological optimization method, which is obtained by repeating the basic negative Poisson's ratio structural unit in two dimensions, as shown in Fig. 2.2 [3]. The structure has good negative Poisson's ratio characteristics, and the structure is simple and easy to control the shape, but its further application is hindered by the limitations in design dimensionality and fabrication. In this paper, the research team of Zheng Dong Ma, has successfully expanded the structure to a macroscopic three-dimensional double-arrow negative Poisson's ratio structure based on the microscopic two-dimensional negative Poisson's ratio structure, as shown in Fig. 2.3a. Compared with the microscopic two-dimensional negative Poisson's ratio structure, the double-arrow negative Poisson's ratio structure has a three-dimensional structure and controllable topological rules, which makes its negative Poisson's ratio characteristics more obvious and easy to manufacture, and greatly expands the application prospects of the negative Poisson's ratio structure, so this paper conducts a more systematic and in-depth study on the double-arrow negative Poisson's ratio structure based on it.

From Fig. 2.3a, it can be seen that the cell structure is assembled by the repetition of a single arrow cell in three directions of space. Therefore, it can be inferred that the design of the arrow cell has an important influence on the performance of the whole cell structure and the negative Poisson's ratio structure. In this paper, the mechanical properties of the double arrow-type negative Poisson's ratio structure are also studied by the method of equivalent mechanical properties. It is assumed that the double arrow-type negative Poisson's ratio structure is an infinite multi-cell structure,

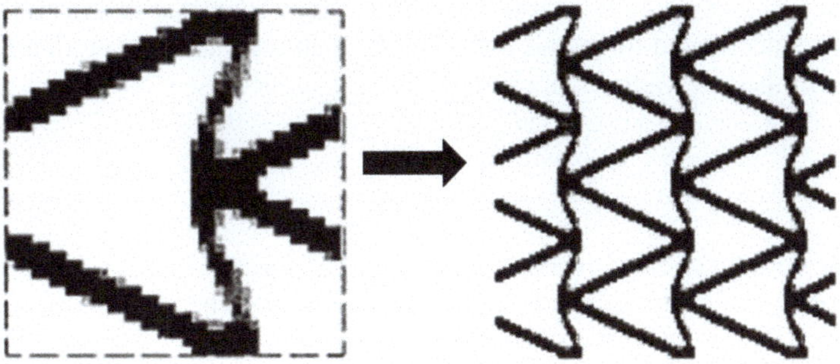

Fig. 2.2 Topology optimization of the resulting negative Poisson's ratio material

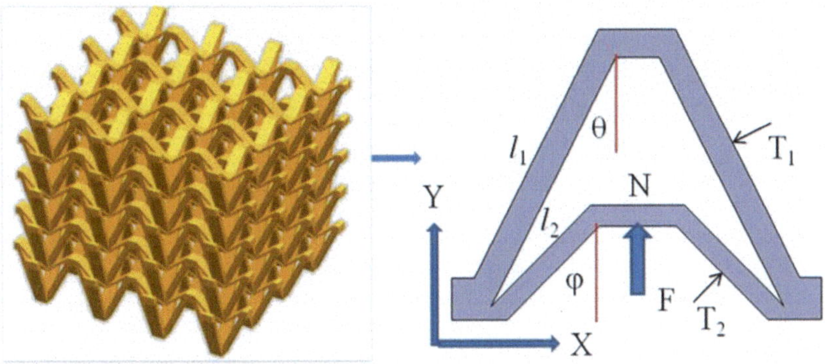

(a) Negative Poisson's ratio structure (b) Negative Poisson's ratio structure cell

Fig. 2.3 Double arrow type negative Poisson's ratio structure

and the arrow-type cell is smaller than the negative Poisson's ratio structure. Then an arrow-type cell is extracted and regarded as continuous and studied by equivalent mechanical properties. The double arrowhead negative Poisson's ratio structure is an anisotropic material, so the equivalent elastic modulus of the structure in a certain direction is the ratio of stress to strain in the direction of the structure under axial compression or tension, and the equivalent Poisson's ratio is the ratio of transverse strain to axial strain of the structure under axial compression or tension.

Figure 2.3b shows the enlarged view of the arrowhead-type single-cell metastable cell, from which it can be seen that the single-cell metastable cell is geometrically symmetric, and the long and short cell metastable walls are the main components of the single-cell metastable cell structure. Assume that l_1 and l_2 are the lengths of the long and short cell walls respectively, and T_1 and T_2 are the thicknesses of the long and short cell walls respectively. θ is the angle between the long cell wall and the symmetry axis of the cell structure, and N is the length of the horizontal wall in the single cell, which is generally set as a constant, and the length of the horizontally connected side of the cell is also fixed as N/2. When subjected to external forces, due to the concentrated force and torque on the cell wall at the same time [4], the deflection of the cell wall can be expressed by Eq. (2.4).

$$d = \frac{Tl^3}{12EI} \tag{2.4}$$

E is the modulus of elasticity coefficient of the basis material and I is the secondary moment of inertia of the cell wall. The external forces in the vertical direction can be expressed as Eqs. (2.5) and (2.6).

$$F_{y_1} = sb(N + l_2 \sin \varphi) \tag{2.5}$$

$$F_{y_2} = sb(N + l_1 \sin \theta) \tag{2.6}$$

s is the stress induced by the external force, such that the deflection on the long cell wall d_1 and the deflection on the short cell wall d_2 can be expressed respectively as

$$d_1 = \frac{s(N + l_2 \sin \varphi) l_1^3 \sin \theta}{ET_1^3} \tag{2.7}$$

$$d_2 = \frac{s(N + l_1 \sin \theta) l_2^3 \sin \varphi}{ET_2^3} \tag{2.8}$$

Since the deformation of the structure in both vertical and horizontal directions is compressive, the strain of the unit cell structure in both directions can be expressed as

$$\varepsilon_y = \frac{d_1 \sin \theta + d_2 \sin \varphi}{l_1 \cos \theta - l_2 \cos \varphi} = \frac{sl_1^2 \sin^2 \theta (N + l_1 \sin \theta)(T_2^3 \sin \varphi + T_1^3 \sin \theta)}{ET_1^3 T_2^3 (\sin \varphi \cos \theta - \sin \theta \cos \varphi)} \tag{2.9}$$

$$\varepsilon_x = -\frac{d_2 \cos \varphi - d_1 \cos \theta}{N + l_1 \sin \theta} = -\frac{sl_1^3 \sin \theta (T_1^3 \sin^2 \theta \cos \varphi - T_2^3 \sin^2 \varphi \cos \theta)}{ET_1^3 T_2^3 \sin^2 \varphi} \tag{2.10}$$

According to the definition of the equivalent properties of the negative Poisson's ratio structure, the equivalent Poisson's ratio and equivalent elastic modulus of the single-cell metastable structure can thus be found, expressed as Eqs. (2.11) and (2.12), respectively.

$$\overline{p} = \frac{\varepsilon_x}{\varepsilon_y} = -\frac{l_1 (T_1^3 \sin^2 \theta \cos \varphi - T_2^3 \sin^2 \varphi \cos \theta)(\sin \varphi \cos \theta - \sin \theta \cos \varphi)}{\sin \theta \sin^2 \varphi (N + l_1 \sin \theta)(T_2^3 \sin \varphi + T_1^3 \sin \theta)} \tag{2.11}$$

$$\overline{E} = \frac{s}{\varepsilon_y} = \frac{ET_1^3 T_2^3 (\sin \varphi \cos \theta - \sin \theta \cos \varphi)}{l_1^2 \sin^2 \theta (N + l_1 \sin \theta)(T_2^3 \sin \varphi + T_1^3 \sin \theta)} \tag{2.12}$$

From Eqs. (2.11) and (2.12), it is known that parameters such as geometry and thickness of the negative Poisson's ratio structure have an important influence on the equivalent Poisson's ratio and equivalent elastic modulus of the structure. The angle between the long cell metric wall and the symmetry axis of the metric structure θ during the design process is generally in the range of $0 < \theta < \varphi < \pi/4$, so it is known from Eq. (2.11) that the structure will always have negative Poisson's ratio structural properties. With the increase of displacement, the formula (2.12) shows that the equivalent elastic modulus of the structure will gradually increase, which means that the negative Poisson's ratio structure has variable stiffness characteristics. The stiffness of the structure will become larger and larger with the action of external force, and the ability of the structure to resist external deformation will become stronger and stronger. In this section, by establishing the relationship between the macro performance mechanical parameters and the microstructure parameters of the negative Poisson's ratio structure, the mechanical properties of the structure with different parameters can be predicted by using the structural equivalent performance mechanical model, which is of great significance for better research and utilization of the negative Poisson's ratio structure. Since the above formula derivation process is based on the small deformation of the structure, it cannot accurately describe the large deformation process of the negative Poisson's ratio structure. In this paper, Sect. 2.3 will study the large deformation characteristics of the negative Poisson's ratio structure by establishing finite element models of different shapes.

2.2.1.2 Inner Concave Hexagonal Negative Poisson's Ratio Structure

In this section, the theoretical analysis of the inner concave hexagonal negative Poisson's ratio structure is carried out based on Gibson's formula and Euler's beam theory. The dimensional parameters of the inner concave hexagonal honeycomb core structure and the established coordinate system are shown in Fig. 2.4a.

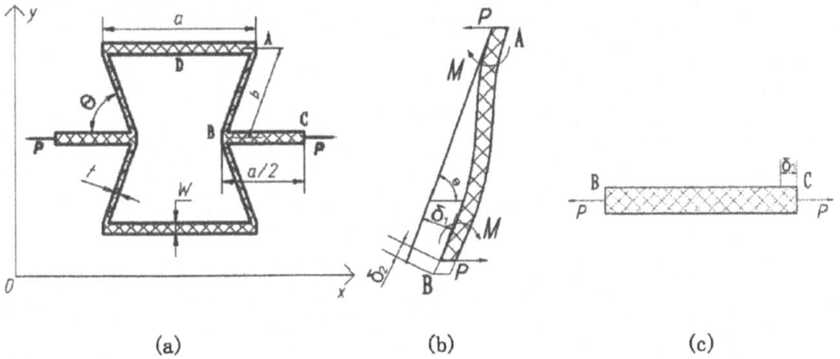

Fig. 2.4 Structure and X-directional deformation of negative Poisson's ratio honeycomb core

When the force P is applied in the x-direction, the deformation produced by the honeycomb core structure is mainly from the deflection of its inclined plate, although the tensile and compressive deformation of its inclined plate and horizontal cross-plate cannot be neglected and is in accordance with the Euler beam theory. The deformation diagram of the inclined plate of the honeycomb core unit body after the force is shown in Fig. 2.6b, and the tensile deformation of its horizontal cross-plate is shown in Fig. 2.6c.

From the equilibrium conditions, the equation between the torque M and the force P can be obtained, and solving the equation yields.

$$M = \frac{1}{2} Pb \sin\theta \tag{2.13}$$

where the force P can be expressed as

$$P = \sigma_1 bh \sin\theta \tag{2.14}$$

h is the width of the honeycomb core unit wall plate, according to the knowledge of material mechanics it is known that the moment of inertia is $I = \frac{ht^3}{12}$, using beam bending theory, the deflection of the honeycomb core wall plate AB after the force is

$$\delta_1 = \frac{Pb^3 \sin\theta}{12E_s I} = \frac{Pb^3 \sin\theta}{E_s ht^3} \tag{2.15}$$

The tensile and compressive deformation of the honeycomb core wall plate is considered non-negligible, and the force P acting on the wall plate AB is caused by the compressive deformation, and the amount of deformation is

$$\delta_2 = \frac{\sigma}{E_s} b = \frac{Pb\cos\theta}{E_s ht} \tag{2.16}$$

The force P acts on the wall plate BC, AD causing it to produce tensile deformation, the elongation is

$$\delta_3 = 2 \times \frac{Pa/2}{E_s wh} = \frac{Pa}{E_s wh} \tag{2.17}$$

The equivalent effect in the x-direction becomes:

$$\varepsilon_1 = \frac{\delta_1 \sin\theta + \delta_2 \cos\theta + \delta_3}{a - b\cos\theta} = \frac{P\sin^2\theta b^3}{E_s ht^3 (a - b\cos\theta)} \left[1 + \left(\mathrm{ctg}^2\theta + \frac{a}{w} \csc^2\theta \frac{t}{b} \right) \frac{t^2}{b^2} \right] \tag{2.18}$$

The equivalent effect in the y-direction becomes:

$$\varepsilon_2 = \frac{\delta_1 \cos\theta - \delta_2 \sin\theta}{b\sin\theta} = \frac{P\cos\theta b^2}{E_s h t^3}\left(1 - \frac{t^2}{b^2}\right) \tag{2.19}$$

Equivalent Poisson's ratio v_{12}^* is

$$v_{12}^* = -\frac{\varepsilon_2}{\varepsilon_1} = -\frac{\cos\theta(a/b - \cos\theta)\left(1 - \frac{t^2}{b^2}\right)}{b\sin^2\theta\left[1 + \left(\mathrm{ctg}^2\theta + \frac{a}{w}\csc^2\theta\frac{t}{b}\right)\frac{t^2}{b^2}\right]} \tag{2.20}$$

The equivalent elastic modulus in the x-direction can be obtained as

$$E_1^* = \frac{\sigma_1}{\varepsilon_1} = \left(\frac{t}{b}\right)^3 \frac{E_s(a/b - \cos\theta)}{\sin^3\theta\left[1 + \left(\mathrm{ctg}^2\theta + \frac{a}{w}\csc^2\theta\frac{t}{b}\right)\frac{t^2}{b^2}\right]} \tag{2.21}$$

δ_2 is the inclined plate compression and δ_3 is the cross-plate elongation.

The equivalent elastic modulus in the x-direction is analyzed above, and the equivalent elastic modulus in the y-direction is analyzed next, as shown in Fig. 2.5a for the dimensional parameters of the honeycomb core structure and its y-direction by the uniform force σ_2. When the y-direction is subjected to the uniform force, the deformation of the honeycomb core structure still mainly comes from the deflection of the inclined plate, and the tensile deformation of the inclined plate also needs to be considered, but the deformation of the horizontal cross-plate in the y-direction is insignificant and can be ignored, so the force and deformation of the inclined plate of the honeycomb core unit is shown in Fig. 2.5b.

Since the honeycomb core cell is subjected to positive stresses in the direction of less σ_2, the bending moment applied to its inclined plate is

$$M = \frac{1}{2} Wb\cos\theta \tag{2.22}$$

Among them

$$W = \sigma_2(a - b\cos\theta)h \tag{2.23}$$

h is the width of the honeycomb core cell wall plate, according to the knowledge of material mechanics can be known moment of inertia is $I = \frac{ht^3}{12}$, using the beam

Bending theory, the deflection of the honeycomb core wall panel AB after the force is

$$\delta_1 = \frac{Wb^3\cos\theta}{12E_s I} = \frac{Wb^3\cos\theta}{E_s h t^3} \tag{2.24}$$

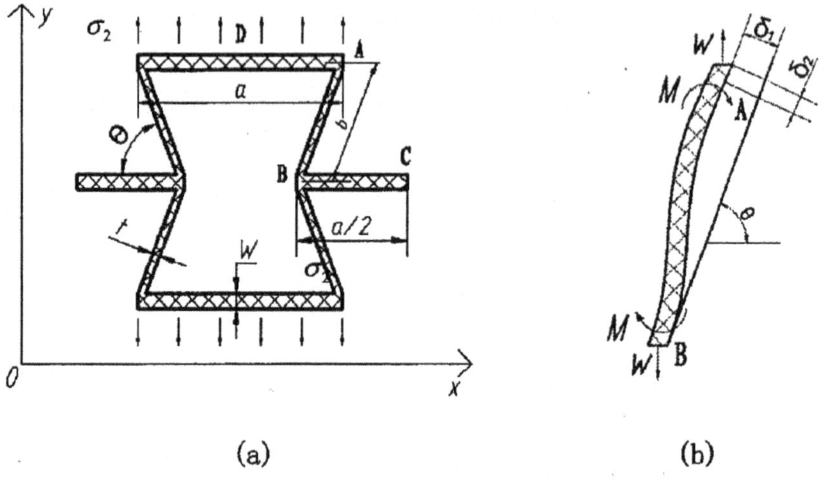

Fig. 2.5 Structure of negative Poisson's ratio honeycomb core and deformation in y-direction

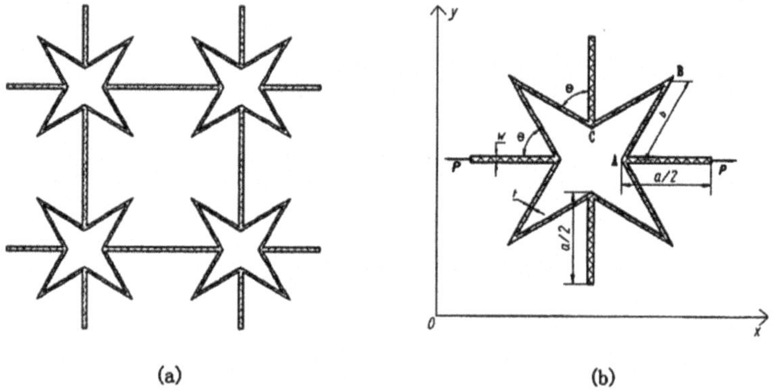

Fig. 2.6 Star structure and structural parameters

Considering the tensile-compressive deformation of the honeycomb core siding as non-negligible, the force W acting on the siding AB causes it to occur.

Tensile deformation with an elongation of

$$\delta_2 = \frac{\sigma}{E_s} b = \frac{W \sin\theta}{E_s h t} \tag{2.25}$$

The equivalent effect in the x-direction becomes:

$$\varepsilon_1 = \frac{\delta_2 \cos\theta - \delta_1 \sin\theta}{a - b\cos\theta} = \frac{W b^3 \sin\theta \cos\theta}{E_s h t^3 (a - b\cos\theta)} \left(1 - \frac{t^2}{b^2}\right) \tag{2.26}$$

The equivalent effect in the y-direction becomes:

$$\varepsilon_2 = \frac{\delta_1 \cos\theta + \delta_2 \sin\theta}{b\sin\theta} = \frac{Wb^2 \sin\theta}{E_s ht^3}\left(\text{ctg}^2\theta + \frac{t^2}{b^2}\right) \tag{2.27}$$

Equivalent Poisson's ratio v_{21}^* is

$$v_{21}^* = -\frac{\varepsilon_1}{\varepsilon_2} = -\frac{b\cos\theta\left(1 - \frac{t^2}{b^2}\right)}{(a/b - \cos\theta)\left(\text{ctg}^2\theta + \frac{t^2}{b^2}\right)} \tag{2.28}$$

Young's modulus in y-direction

$$E_2^* = \frac{\sigma_2}{\varepsilon_2} = \left(\frac{t}{b}\right)^3 \frac{E_s}{\sin\theta(a/b - \cos\theta)\left(\text{ctg}^2\theta + t^2/b^2\right)} \tag{2.29}$$

2.2.1.3 Star-Shaped Negative Poisson's Ratio Structure

The array-type structure of the star-shaped honeycomb core structure is shown in Fig. 2.6a. Given that the star-shaped structure has the same mechanical properties in x and y directions, only one of its directions needs to be analyzed, and its cell parameters and coordinate system are shown in Fig. 2.6b.

When its x-direction is acted by the force P, the deflection deformation produced by its unit wall plate is shown in Fig. 2.7a, b, based on Euler beam theory, without considering the expansion deformation of the wall plate.

For the inclined plate AB, its force and deformation are shown in Fig. 2.7a, by the equilibrium condition, the equation between the torque M and the force P, solve the equation to get

$$M = \frac{1}{2} Pb\sin\theta_1 \tag{2.30}$$

where the force P can be expressed as

$$P = \sigma_1 bh\sin\theta_1 \tag{2.31}$$

h is the width of the honeycomb core unit wall plate, according to the knowledge of material mechanics it is known that the moment of inertia is $I = \frac{ht^3}{12}$, using beam bending theory, the deflection of the honeycomb core wall plate AB after the force is

$$\delta_1 = \frac{F_1 b^3}{3EI} - \frac{Mb^2}{2EI} = \frac{pb^3 \sin\theta_1}{3EI} - \frac{pb^3 \sin\theta_1}{4EI} = \frac{pb^3 \sin\theta_1}{Eht^3} \tag{2.32}$$

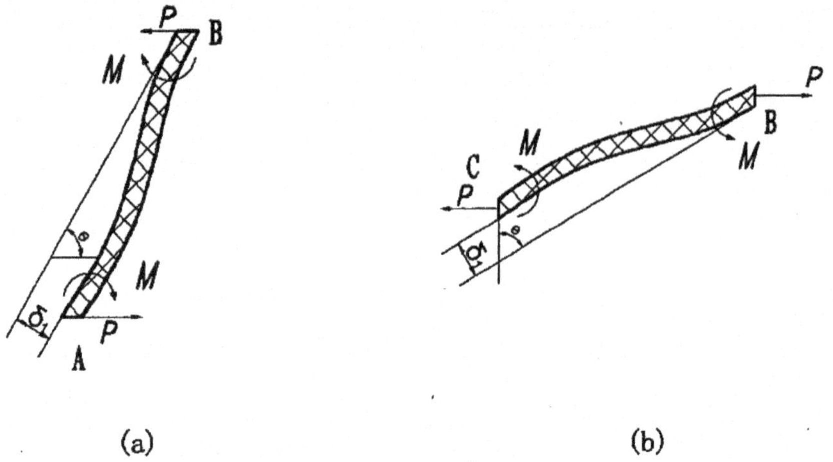

Fig. 2.7 Star structure work direction force deformation

For the inclined plate BC, the force and deformation are shown in Fig. 2.9b, and by the equilibrium condition, the equation between the torque M and the force P can be obtained, and solving the equation yields.

$$M = \frac{1}{2} Pb \cos\theta_1 \tag{2.33}$$

The deflection of the honeycomb core wall panel BC after the force is

$$\delta_2 = \frac{F_1 b^3}{3EI} - \frac{Mb^2}{2EI} = \frac{pb^3 \cos\theta_1}{3EI} - \frac{pb^3 \cos\theta_1}{4EI} = \frac{pb^3 \cos\theta_1}{Eht^3} \tag{2.34}$$

By the amount of deformation produced by the force and the effective length of the unit body, the equivalent effect in the x-direction becomes

$$\varepsilon_x = \frac{\delta_x}{l_x} = \frac{\delta_1 \sin\theta_1 + \delta_2 \cos\theta_1}{\frac{a}{2} + b(\sin\theta_1 - \cos\theta_1)} = \frac{2Pb^3}{Eht^3[a + 2b(\sin\theta_1 - \cos\theta_1)]} \tag{2.35}$$

By the amount of deformation produced by the force and the effective length of the unit body, the equivalent effect in the y-direction becomes

$$\varepsilon_y = \frac{\delta_y}{l_y} = \frac{\delta_1 \cos\theta_1 + \delta_2 \sin\theta_1}{\frac{a}{2} + b(\sin\theta_1 - \cos\theta_1)} = \frac{4Pb^3 \sin\theta_1 \cos\theta_1}{Eht^3[a + 2b(\sin\theta_1 - \cos\theta_1)]} \tag{2.36}$$

Then the equivalent Poisson's ratio of the star structure is

$$\nu_{12} = -\frac{\varepsilon_y}{\varepsilon_x} = -2\sin\theta_1 \cos\theta_1 \tag{2.37}$$

The equivalent elastic modulus in the x-direction is

$$E_1 = \frac{\sigma_1}{\varepsilon_x} = E_s \frac{t^3[a + 2b(\sin\theta_1 - \cos\theta_1)]}{2b^4 \sin\theta_1} \tag{2.38}$$

2.2.1.4 Chiral Negative Poisson's Ratio Structure

Chiral honeycomb structures are obtained by the energy principle for the in-plane modulus of elasticity of their structures. According to the principle of energy conservation, the strain energy due to the strain in the direction of i is equal to the sum of the energy stored in each bent ligament in the cell per unit volume (W_{rib}), i.e.

$$U = \frac{1}{2} E_i \varepsilon_i^2 = \frac{n}{V} W_{rib} \tag{2.39}$$

where

- U is the strain energy of the representative cell.
- n is the number of ligaments of the taken unit.
- V is the volume of the representative body element.

Since it is assumed that only bending deformation of the ligament occurs, only the work done by the bending moment is available. According to standard beam theory, the energy stored in a single ligament is equal to the work done by the bending moment M at both ends of the ligament, i.e.

$$W_{rib} = 2 \int_0^{\varphi} M d\varphi \tag{2.40}$$

where

- ϕ is the angle at which the ends of the ligament turn over due to the bending moment M.

The four-ligament reverse chiral honeycomb structure is affected by the ligament length in x and y directions, so its Poisson's ratio will change, and also in these two directions, the Young's modulus is also different. As (Fig. 2.8).

Before calculating the in-plane Young's modulus, the effect of the thickness of the ligament and cylindrical junction of the chiral honeycomb structure was noted.

Loudness, the equivalent ligament length is used, and the equation for calculating the equivalent ligament length is obtained by a simple geometric relationship.

$$L_{eq} = L - 2\sqrt{r^2 - (r - t)^2} \tag{2.41}$$

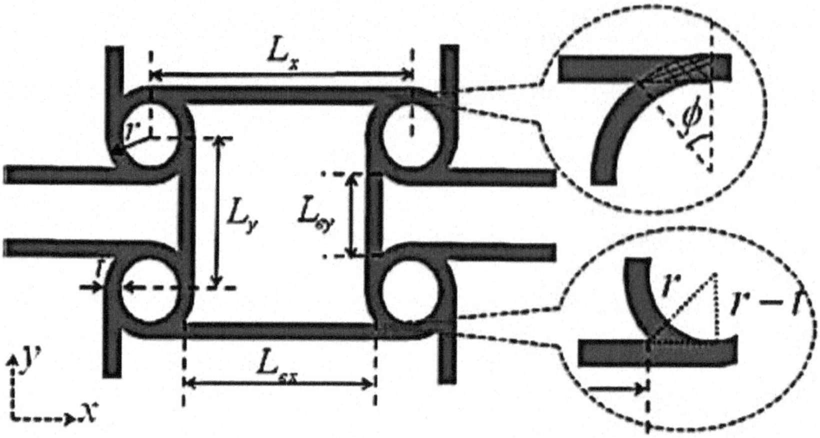

Fig. 2.8 Schematic diagram of the structural parameters of the four-ligament reverse chiral fovea

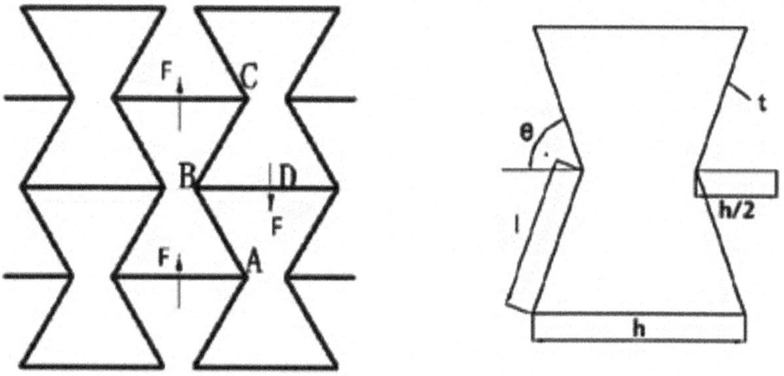

Fig. 2.9 Structural parameters

So for the four ligament reverse chiral foveal structures there are.

$$\begin{aligned} L_{ex} &= L_x - 2\sqrt{r^2 - (r-t)^2} \\ L_{ey} &= L_y - 2\sqrt{r^2 - (r-t)^2} \end{aligned} \tag{2.42}$$

The angle of rotation of the cylinder is obtained from the disturbance equation of the beam as

$$\phi = ML_e/2E_sI \tag{2.43}$$

where

I is the moment of inertia of the ligament, $I = \frac{bt^3}{12}$.

The energy stored in the ligament is.

$$W_{nb} = \frac{E_s bt^3 \phi^2}{6L_e} \tag{2.44}$$

Substituting the formula, the following equation is obtained.

$$\begin{aligned}
\frac{1}{2}E_x\varepsilon_x^2 &= 4\left[\frac{E_s bt^3 L_x^2 \varepsilon_x^2}{24L_{ex}(r-t/2)^2} + \frac{E_s bt^3 L_x^2 \varepsilon_x^2}{24L_{ey}(r-t/2)^2}\right]\frac{1}{4L_xL_yb} \\
\frac{1}{2}E_x\varepsilon_x^2 &= 4\left[\frac{E_s bt^3 L_y^2 \varepsilon_y^2}{24L_{ex}(r-t/2)^2} + \frac{E_s bt^3 L_y^2 \varepsilon_y^2}{24L_{ey}(r-t/2)^2}\right]\frac{1}{4L_xL_yb}
\end{aligned} \tag{2.45}$$

The above two equations are reduced to dimensionless form

$$\begin{aligned}
E_x &= \frac{E_s\beta^3\alpha_x}{12(1-\beta/2)^2\alpha_y}\left(\frac{1}{\alpha_x - 2\sqrt{2\beta-\beta^2}} + \frac{1}{\alpha_y - 2\sqrt{2\beta-\beta^2}}\right) \\
E_y &= \frac{E_s\beta^3\alpha_y}{12(1-\beta/2)^2\alpha_x}\left(\frac{1}{\alpha_x - 2\sqrt{2\beta-\beta^2}} + \frac{1}{\alpha_y - 2\sqrt{2\beta-\beta^2}}\right)
\end{aligned} \tag{2.46}$$

where E_S is the Young's modulus of the material used to make the honeycomb structure.

Equations (2.45) and (2.46) are the analytical expressions for the elastic constants of the anisotropy of the reverse chiral foveal structure of the four ligaments. In particular, when the ligament length $L_x = L_y$, i.e., $\alpha_x = \alpha_y = \alpha$, there is

$$E_x = E_y = \frac{E_s\beta^3}{6(1-\beta/2)^2}\frac{1}{\alpha - 2\sqrt{2\beta-\beta^2}} \tag{2.47}$$

At this time, the in-plane Young's modulus of the honeycomb structure is equal, and it can become an in-plane isotropic structure.

2.2.2 *Negative Poisson's Ratio Structural Shear Modulus*

It is required that the honeycomb core structure has sufficient strength and stiffness in this direction; therefore, it is necessary to conduct a theoretical analysis of the equivalent shear modulus of the structure to lay a theoretical foundation for the design and optimization of the honeycomb core structure.

2.2.2.1 Shear Modulus Within the Inner Concave Hexagonal Honeycomb SurfaceG_{12}

As shown in Fig. 2.9, when the negative Poisson's ratio structure is subjected to a shear force F in the y-direction, its deformation is shown in Fig. 2.10. According to the knowledge of material mechanics, the combined moment acting on point B is equal to the sum of the moments acting on the BC and BA sides

$$M = \frac{1}{4}Fh + \frac{1}{4}Fh = \frac{1}{2}Fh \tag{2.48}$$

where F is the shear force acting in the y-direction of the negative Poisson's ratio structure, h is length of cross-ribbed plate.

According to the knowledge of material mechanics, it is known that the deformation of point D δ

$$G = \frac{Ml^2}{6E^*I} \tag{2.49}$$

where

l length of slant rib plate.
I cross-sectional moment of inertia of the cross-wall plate.
E^* material modulus of elasticity.

It follows that the angle of twist of the wall plate BD is

$$\varphi = \frac{Fhl}{24E^*I} \tag{2.50}$$

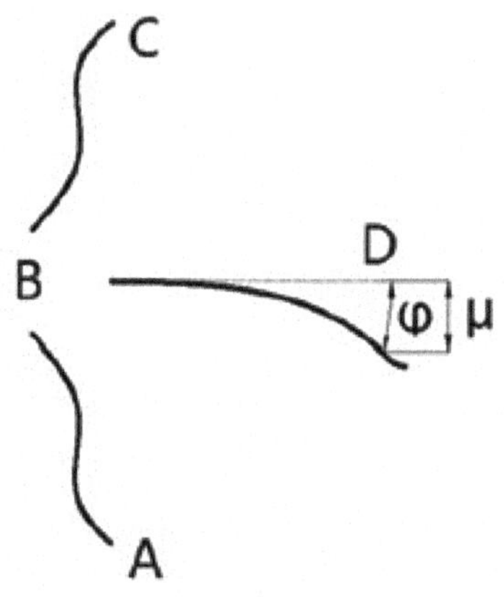

Fig. 2.10 Shear deformation in the y-direction of the internal concave hexagonal negative Poisson's ratio structure

Under the action of shear force F, the deformation of point D in the y-direction about point B μ is

$$\mu = \frac{1}{2}\varphi h + \frac{F}{3E^{*}I}\left(\frac{h}{2}\right)^{3} = \frac{Fh^{2}}{48E^{*}I}(l + 2h) \tag{2.51}$$

Shear strain γ is

$$\gamma = \frac{\mu}{h/2} = \frac{Fh}{24E^{*}I^{*}}(l + 2h) \tag{2.52}$$

Shear stress τ is

$$\tau = \frac{F}{2bl\sin\theta} \tag{2.53}$$

Then the in-plane shear modulus G_{12} is

$$G_{12} = \frac{\tau}{\gamma} = \frac{12E^{*}I}{hbl(l + 2h)\sin\theta} = \frac{E^{*}t^{3}}{lh(l + 2h)\sin\theta} \tag{2.54}$$

The above equation is the in-plane shear modulus of the internal concave hexagonal negative Poisson's ratio structure G_{12}, which provides the theoretical basis for the optimal design of the internal concave hexagonal negative Poisson's ratio structure later.

2.2.2.2 Shear Modulus Outside the Inner Concave Hexagonal SurfaceG_{13}

The analysis of the out-of-plane shear modulus of the inwardly concave hexagonal negative Poisson's ratio structure is complicated by the nonlinear deformation of the core wall due to the action of the surrounding core wall during out-of-plane shear, which results in a complex distribution of stresses in the inwardly concave hexagonal negative Poisson's ratio structure. Only numerical simulations can be used for accurate analysis, but the upper and lower limits of the equivalent out-of-plane shear modulus G_{13} and G_{23} can be predicted using the principle of minimum potential energy and the principle of minimum residual energy. The upper limit of the equivalent out-of-plane shear modulus G_{13} and G_{23} can be obtained by applying the principle of minimum potential energy, and the lower limit of the equivalent out-of-plane shear modulus and can be obtained by applying the principle of minimum residual energy. The minimum potential energy principle is expressed as follows: among all possible displacement fields satisfying the displacement coordination condition, the solution corresponding to the minimum structural strain energy is the true solution; the minimum residual energy principle is expressed as follows: among all possible

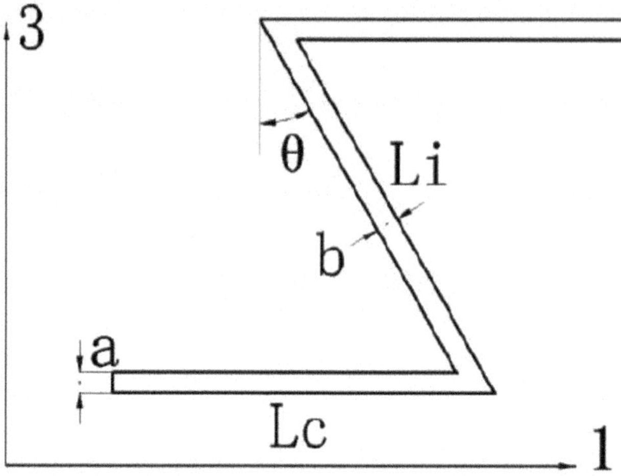

Fig. 2.11 Simplified characterization unit

stress fields satisfying the force equilibrium condition, the stress field corresponding to the minimum strain energy is the true solution (Fig. 2.11).

When the inwardly concave hexagonal type of negative Poisson's ratio single cell structure produces uniform shear strain γ_{13}, it is known from the strain coordination condition

$$\begin{cases} \gamma_h = 0 \\ \gamma_l = \gamma_{13} cos\theta \end{cases} \tag{2.55}$$

The volume of the single cell structure is

$$V = hl\cos\theta b \tag{2.56}$$

where

b wall plate height.

The strain energy of each part of the single cell is

$$\begin{cases} u_c = 0 \\ u_i = \frac{1}{2}G_s\gamma_i^2 = \frac{1}{2}G_s\gamma_{13}^2\cos^2\theta \end{cases} \tag{2.57}$$

where

u_c cross-wall plate strain energy.
u_i Strain energy of inclined wall plate.

From the principle of minimum potential energy, we know that

$$\frac{1}{2}G_{13}\gamma_{13}^2 V \leq \frac{1}{2}\sum_{i=1}^{n}(G_s\gamma_i^2 V_i) \tag{2.58}$$

$$G_{13} \leq \frac{t\cos\theta}{h}G_s \tag{2.59}$$

When the inner concave hexagonal type negative Poisson's ratio single cell structure is subjected to uniform shear stress τ_{13}, by the principle of minimum potential energy

$$\frac{1}{2}\frac{\tau_{13}^2}{G_{13}}V \leq \frac{1}{2}\sum_{i=1}^{n}\left(\frac{\tau_i^2}{G_s}V_i\right) \tag{2.60}$$

From the principle of equilibrium, we know that

$$\tau_{13}hl\cos\theta = \tau_b\cos\theta tl \tag{2.61}$$

Substituting into Eq. (2.12), we can see that

$$G_{13} \geq \frac{t\cos\theta}{h}G_s \tag{2.62}$$

In summary, it can be seen that

$$G_{13} = \frac{t\cos\theta}{h}G_s \tag{2.63}$$

The above equation is the out-of-plane shear modulus G_{13} for the in-concave hexagonal negative Poisson's ratio structure, which is the later in-concave hexagonal. The optimal design of the negative Poisson's ratio structure provides a theoretical basis.

2.2.2.3 External Shear Modulus of the Inner Concave Hexagonal SurfaceG_{23}

When the inner concave hexagonal type negative Poisson's ratio single cell structure produces uniform shear strain γ_{23}, it is known from the strain coordination condition

$$\begin{cases} \gamma_h = \gamma_{23} \\ \gamma_l = \gamma_{23}\sin\theta \end{cases} \tag{2.64}$$

The volume of the inner concave hexagonal type negative Poisson's ratio single cell structure is

$$V = hl\cos\theta b \tag{2.65}$$

The strain energy of each part of the negative Poisson's ratio single cell structure of the inner concave hexagonal type

$$\begin{cases} u_h = \frac{1}{2}G_s\gamma_{23}^2 \\ u_l = \frac{1}{2}G_s\gamma_{23}^2\sin^2\theta \end{cases} \tag{2.66}$$

From the principle of minimum potential energy, we know that

$$\frac{1}{2}G_{23}Y_{23}^2V \leq \frac{1}{2}\sum_{i=1}^{n}(G_sY_i^2V_i) \tag{2.67}$$

$$G_{23} \leq \frac{t(h + l\sin^2\theta)}{hl\cos\theta}G \tag{2.68}$$

When the inner concave hexagonal type negative Poisson's ratio single cell structure is subjected to uniform shear stress τ_{23}, it is known from the principle of minimum potential energy

$$\frac{1}{2}\frac{\tau_{23}^2}{G_{23}}V \leq \frac{1}{2}\sum_{i=1}^{n}\left(\frac{\tau_i^2}{G_s}V_i\right) \tag{2.69}$$

From the principle of force equilibrium, we know that

$$\tau_{23}hl\cos\theta = \tau_a th + \tau_b lt\sin\theta \tag{2.70}$$

Substituting into Eq. (2.20), we can see that

$$G_{23} \geq \frac{t(h + l\sin\theta)^2}{hl(h + l)\cos\theta}G \tag{2.71}$$

In summary, it can be seen that

$$\frac{t(h + l\sin^2\theta)}{hl\cos\theta}G_s \geq G_{23} \geq \frac{t(h + l\sin\theta)^2}{hl(h + l)\cos\theta}G \tag{2.72}$$

The above equation is the upper and lower limits of the out-of-plane shear modulus G_{23} for the inner concave hexagonal negative Poisson's ratio structure, which provides the theoretical basis for the optimal design of the inner concave hexagonal negative Poisson's ratio structure later.

2.2.3 Relative Density of Negative Poisson's Ratio Structure

The application of the basis material of the negative Poisson's ratio structure has an important influence on the mechanical properties of the structure. The shape and structure of the basis material and the microstructure determine the macroscopic properties of the negative Poisson's ratio structure. In order to keep consistent with the general research methods of mechanical properties of materials, people generally regard multi-cell structures such as negative Poisson's ratio structures as continuous media and describe them in the way of equivalent performance and relative performance when studying multi-cell structures such as negative Poisson's ratio structures. The main mechanical evaluation indexes commonly used for equivalent performance are equivalent elastic modulus and equivalent Poisson's ratio. The equivalent performance is an important research index of the performance of the multi-cell structure, and the ratio of the equivalent performance of the multi-cell structure such as the negative Poisson's ratio structure to the performance of the basis material during the research process is called the relative performance of the multi-cell structure [5]. The relative performance also has an important influence on the equivalent performance of the negative Poisson's ratio structure. The main mechanical evaluation indexes of the relative performance mainly include relative elastic modulus and relative density.

In the study of negative Poisson's ratio structure, the relative density of the structure is the ratio of the density of the negative Poisson's ratio structure to the density of the basis material, and the relative density is an important performance index to describe the properties of the negative Poisson's ratio structure [6]. Due to the honeycomb structural properties of the negative Poisson's ratio structure, the relative density can be taken in the range between 0 and 1 according to the definition of relative density. When the gap between the cells of the structure is small and the cell wall is filled with the basis material, the density of the negative Poisson's ratio structure will tend to be the density of the basis material, and the value will be close to 1. When the gap between the cells of the negative Poisson's ratio structure is large and there is basically no filling material between the structure, the relative density of the negative Poisson's ratio structure will tend to be close to 0. The size of the cells in the material is not directly related to the relative density, and the relative density mainly depends on the structure of the cells. Therefore, it is important to establish the mechanical model of the shape parameters and relative density of the negative Poisson's ratio structure cell.

The schematic diagram of the meta structure of the double-arrow-shaped negative Poisson's ratio structure studied in this paper is shown in Fig. 2.12. In the calculation of the relative density of the cell, the smallest repeating unit of the structure can be considered as a complete calculation unit, and the calculation unit taken in the analysis of the relative density of the double-arrow-shaped negative Poisson's ratio structure cell in this paper is shown in the shaded part in Fig. 2.4. Assuming that ρ_R is the relative density of the structure, ρ_N is the equivalent density of the material, and ρ_M is the density of the base material, we therefore have

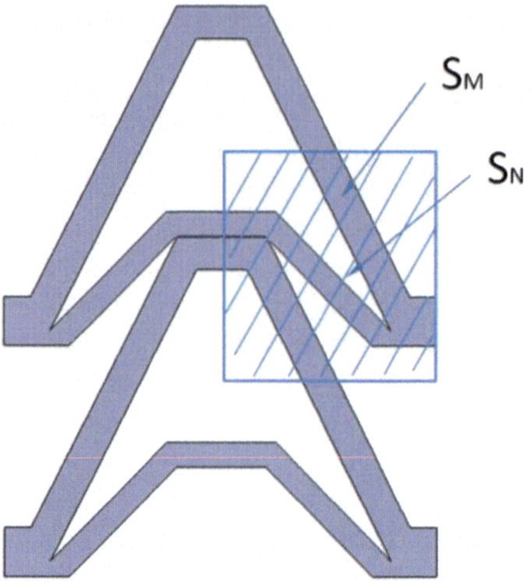

Fig. 2.12 Schematic diagram of the relative density of two-dimensional arrow-shaped meta cells

$$\rho_R = \frac{\rho_N}{\rho_M} \tag{2.73}$$

Assuming that S_N is the area of the computational cell and S_M is the area of the substrate material in the computational cell, it follows from the definition of the equivalent density of the negative Poisson's ratio structure that

$$\rho_N = \frac{S_M a \rho_M}{S_N a} \tag{2.74}$$

where, assuming the same width of the long and short cell walls in the cell, a is the width of the cell wall in the calculated cell. According to the parameter schematic of the double-arrow type cell structure in Fig. 2.3b, the relationship between S_N and S_M and the cell shape parameters of the negative Poisson's ratio structure can be found as follows, respectively.

$$S_M = T_1 l_1 + T_2 l_2 + 2 * T_1 * \frac{1}{2} N + 2 * T_2 * \frac{1}{2} N = T_1(l_1 + N) + T_2(l_2 + N) \tag{2.75}$$

$$S_N = (l_1 \sin\theta + N)(l_1 \cos\theta - l_2 \cos\varphi) \tag{2.76}$$

Therefore, combining the above expressions, the relationship between the relative density of the material and the metameric cell shape parameters can be derived as follows.

$$\rho_R = \frac{T_1(l_1 + N) + T_2(l_2 + N)}{(l_1 \sin\theta + N)(l_1 \cos\theta - l_2 \cos\varphi)} \tag{2.77}$$

When the negative Poisson's ratio structure is subjected to external forces, the structure will deform in both horizontal and vertical directions, and the relative density of the material will change. Assume that ε_t is the instantaneous strain of the calculated unit of the negative Poisson's ratio structure, so the instantaneous relative density of the material can be defined as

$$\rho_R(t) = \rho_R \frac{1}{1 - \varepsilon_t} \tag{2.78}$$

When the deformation of the negative Poisson's ratio structure gradually becomes larger because of the action of external forces, the relative density of the negative Poisson's ratio structure will gradually increase according to Eq. (2.77), and theoretically, when all the voids of the negative Poisson's ratio structure are filled, the relative density of the structure will tend to be similar to the density of the basis material.

By establishing the mechanical model between the relative density of the negative Poisson's ratio structure and the thickness of the cell wall, the angle between the cell walls and the thickness of the cell wall, the relative density of the material can be obtained according to the shape parameters of the cell in the design process of the microstructure of the negative Poisson's ratio structure, so as to provide guidance for the design of the structure.

2.3 Study of Static Properties of Negative Poisson's Ratio Structure

In order to study the static properties of the negative Poisson's ratio structure, this paper establishes a finite element model of the 4 $\times$ 4 unit negative Poisson's ratio structure based on hexahedral cells. The external force of size 50 N acts on the upper part of the structure in the process of analysis, except for the 5 degrees of freedom in the horizontal direction at the bottom of the restrained structure. Figure 2.13a, b show the finite element model and stress cloud of the negative Poisson's ratio structure, respectively. By comparing Fig. 2.13a, b, it can be seen that the negative Poisson's ratio structure is displaced in compression in both horizontal and vertical directions, so it is known that the structure has negative Poisson's ratio structural properties and the density of the negative Poisson's ratio structure will gradually increase with the loading of the force.

To comparatively study the static properties of the negative Poisson's ratio structure, five different structures were obtained by changing the geometry and thickness parameters of the negative Poisson's ratio structure, etc. The design parameters of the five different structures are shown in Table 2.1, and the finite element models of the structures were established and the same constraints and external forces were applied. For the study of the structures, the upper symmetric center point of the structure is used as the reference point, and the mechanical performance evaluation index

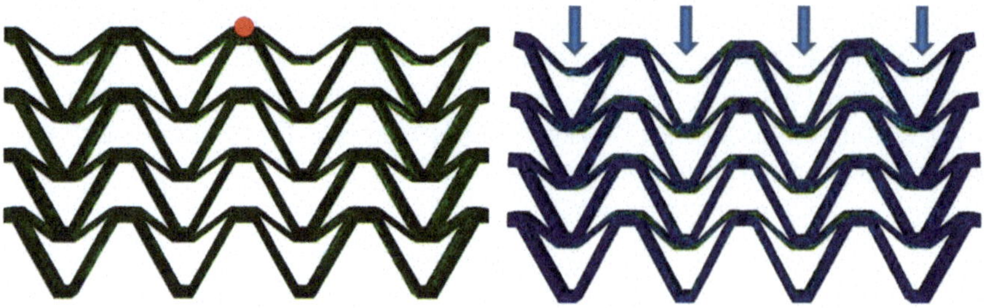

(a) Double-arrow negative Poisson's ratio structure finite element model

(b) Double-arrow negative Poisson's ratio structure stress cloud

Fig. 2.13 Finite element model and deformation of the structure with negative Poisson's ratio

of the negative Poisson's ratio structure is calculated by calculating the displacement of the reference point of the structure, etc. The reference point is shown as the dot in Fig. 2.13a.

Figure 2.14 shows the comparison of force–displacement curves for five different structures.

Table 2.1 Design parameters for different negative Poisson's ratio structures

	θ/rad	φ/rad	T_1/mm	T_2/mm
Structure A	0.3	0.6	2	1
Structure B	0.4	0.6	2	1
Structure C	0.3	0.7	2	1
Structure D	0.3	0.6	1.8	1
Structure E	0.3	0.6	2	1.5

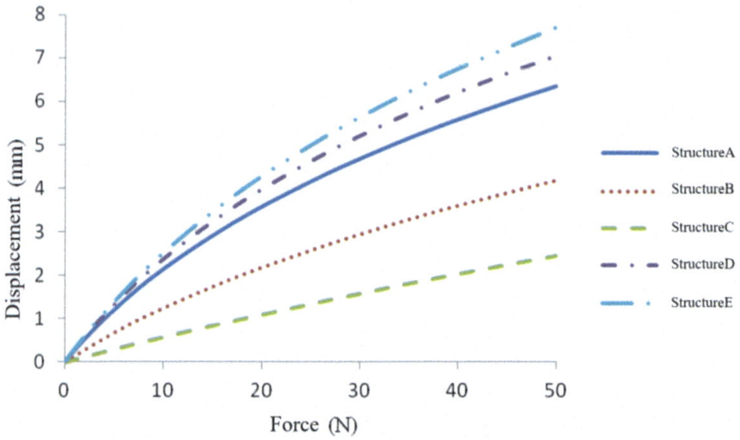

Fig. 2.14 Force–displacement curves for structures with different negative Poisson's ratios

Figure 2.15 shows the displacement of the reference point in structure C is the smallest, and the reference point in structure E always has the largest displacement. When the displacement of the structure is small, the displacement of the reference point in structure E is basically the same as that in structure A and structure D. However, when the deformation of the structure gradually increases, the displacement of the reference point in structure E grows the fastest and the final deformation is the largest. As the force increases, all the curves gradually become flat, that is, the efficiency of the curve becomes smaller and smaller. It can be seen that as the force is loaded, the stiffness of the negative Poisson's ratio structure gradually increases. Based on the above analysis, it can be seen that the double-arrow negative Poisson's ratio structure has variable stiffness characteristics, and the stiffness of the structure will increase with the increase of compression deformation. At the same time, the negative Poisson's ratio structure with different design parameters, even if it has the same initial stiffness, the change of structural stiffness is also different due to different design parameters.

Figure 2.15 shows the equivalent Poisson's ratio-force curves for five different structures. From the figure, it can be seen that the equivalent Poisson's ratios of all structures are between 0 and -1. Structure E has the smallest Poisson's ratio, while structure C has the largest Poisson's ratio and changes the least during loading. With the loading of force, the Poisson's ratio of all structures becomes smaller and smaller, and the slope of each curve is different in the process of change, i.e., the trend of change is different for different structures. The above analysis shows that the negative Poisson's ratio structures with different design parameters have different sizes of negative Poisson's ratio values between 0 and -1, and the Poisson's ratio of the structure gradually becomes smaller with the increase of loading, and the change trend of the negative Poisson's ratio structures with different design parameters is different.

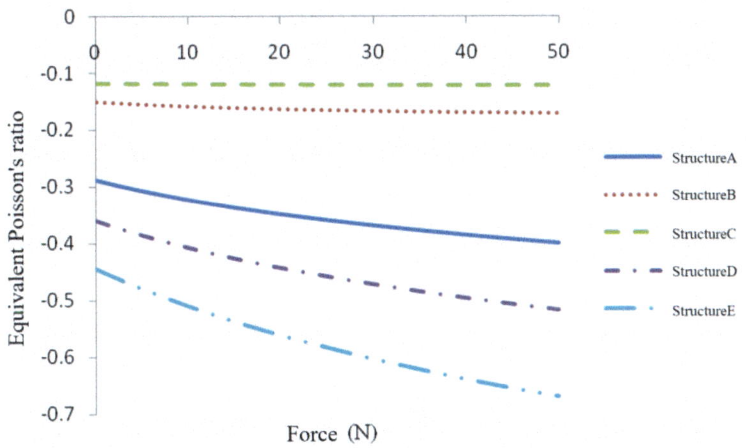

Fig. 2.15 Equivalent Poisson's ratio-force curves for different negative Poisson's ratio structures

Figure 2.16 shows the equivalent elastic modulus-force curves of different structures. From the figure, it can be seen that the equivalent elastic modulus of different structures increases with the loading of force, and the rate of increase varies for the negative Poisson's ratio structures with different design parameters. Structure B has the smallest equivalent elastic modulus among all structures. Although the equivalent elastic modulus of structure E is smaller than that of structure C in the initial stage of force loading, the growth rate of structure E is larger as the deformation increases, and the elastic model moduli of both are equal during the loading process, and finally the equivalent elastic modulus of structure E is larger than that of structure C. Comprehensive analysis shows that the negative Poisson's ratio structures with different design parameters have different equivalent elastic modulus, and the equivalent elastic modulus of the structure gradually increases with the increase of deformation, and the change trend is related to the design parameters of the structure. This special performance makes the negative Poisson's ratio structure can be widely used in the design of automobile suspension and other structures, so that the application of negative Poisson's ratio structure should be designed according to the actual needs of the relevant parameters.

From the above analysis, it can be seen that after changing the microscopic topology of the negative Poisson's ratio structure, the equivalent elastic modulus and equivalent Poisson's ratio of different negative Poisson's ratio structures vary with the force, but these structures have negative Poisson's ratio characteristics and the ability of the structure to resist external deformation increases with the increase of deformation. In this section, the static characteristics of the negative Poisson's ratio structure are studied, which is of great significance for understanding the special mechanical properties of the negative Poisson's ratio structure and promoting the further research and application of the negative Poisson 's ratio structure.

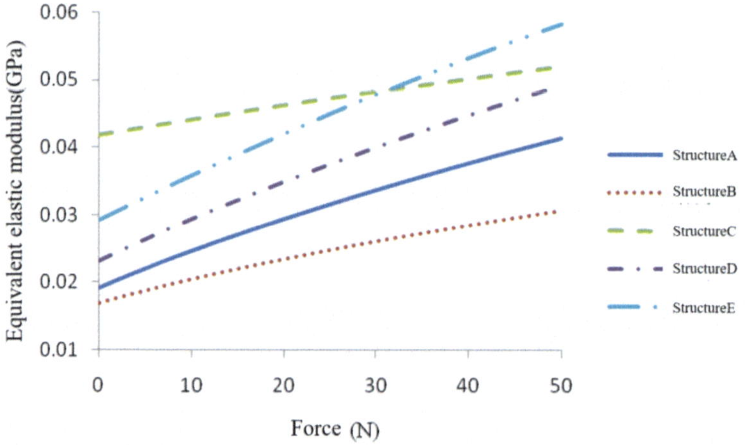

Fig. 2.16 Equivalent elastic modulus-force curves for structures with different negative Poisson's ratios

2.4 Study of Energy Absorption Characteristics of Negative Poisson's Ratio Structure

2.4.1 Negative Poisson's Ratio Structure Compression Properties

Due to the special concave structure, the negative Poisson's ratio structure can have good energy absorption characteristics and impact resistance while maintaining a lighter structural mass, and its special energy absorption characteristics can be well applied to the design of energy absorbing structures [7]. Therefore, it is important to master the in-plane compression characteristics and energy absorption mechanism of negative Poisson's ratio structure to study the deformation law of the structure under impact load and the energy absorption law of negative Poisson's ratio material, which will promote the better application of negative Poisson's ratio structure in engineering design.

When the negative Poisson's ratio structure is subjected to external load, the load is in-plane when it is in the same plane with the cross-section of the structure, and out-of-plane when they are not in the same plane, so the macroscopic mechanical properties of the negative Poisson's ratio structure also include in-plane and out-of-plane properties, and the elastic properties of the negative Poisson's ratio structure are generally anisotropic when it is in-plane. In the study of multi-cell structures such as negative Poisson's ratio structures, the whole structure is generally considered as a continuous material. For multi-cell materials, the in-plane mechanical properties of the material are very closely related to its overall performance and affect its application in engineering. The negative Poisson's ratio structure belongs to the generalized honeycomb structure, so it has commonality with the in-plane properties of the general honeycomb structure, but its in-plane compression process is different from the general conventional honeycomb structure because of its different microscopic topology. The mechanical properties of the negative Poisson's ratio structure are not only influenced by the basis material, but also closely related to the shape and distribution of the negative Poisson's ratio structural cells. Figure 2.17 shows the comparison curves of the in-plane compression properties of the negative Poisson's ratio structure and the general honeycomb structure.

Assuming that the basis material is elastic–plastic material, the curve a in Fig. 2.17 shows the one-way compressive stress–strain curve of a general honeycomb material. It can be seen from the figure that the curve a contains mainly 3 different parts, the first part is the linear elastic phase, within which the cell wall of the microstructure undergoes bending deformation, and the bending deformation is the main deformation of the material in this phase, which is generally described by the slope of the curve, i.e., the equivalent elastic modulus of the material; the second part is the stress plateau phase, within which the deformation of the material gradually increases, while the stress magnitude remains almost constant in size. The deformation in this phase varies for different basis materials and mainly includes plastic collapse, elastic

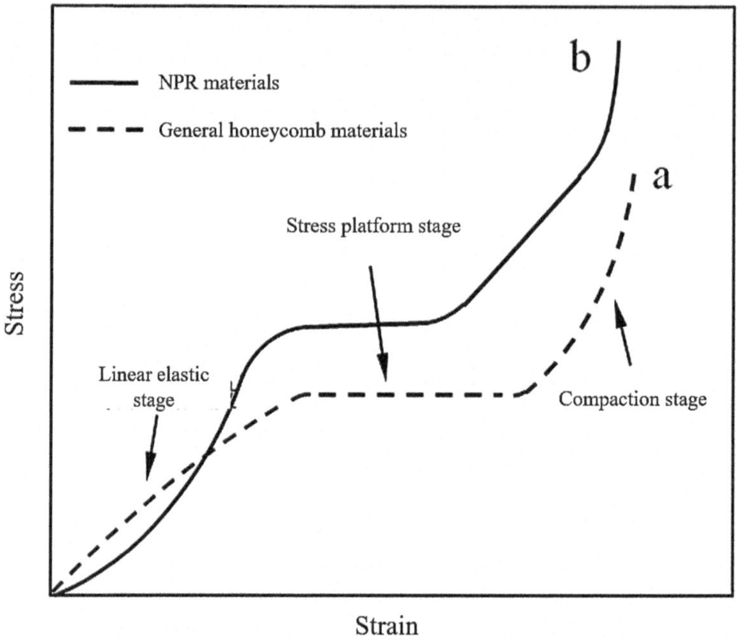

Fig. 2.17 Unidirectional compressive stress–strain curves for honeycomb and negative Poisson's ratio materials

buckling and brittle fracture, etc. For elastoplastic materials, plastic collapse generally occurs in this phase, and the microstructure of the material collapses layer by layer during compression until all cells collapse. The stress of plastic collapse determines the magnitude of stress in the region, and the deformation of the microstructure in the stress plateau stage has an important effect on the macroscopic mechanical properties of the material. The third part is the densification stage, after the collapse of all the cells, the cell walls of the cells are contacted and laminated together, and the stress of the material rises rapidly, and the equivalent elastic modulus will tend to the elastic modulus of the basis material, and the corresponding strain magnitude of the material is important for the energy absorption of the honeycomb structure when the densification of the material just occurs in this stage.

The curve b in Fig. 2.17 shows the unidirectional compression stress–strain curve of the negative Poisson's ratio material. It can be seen that the unidirectional compression stress–strain curve of negative Poisson's ratio material is similar to the stress–strain curve of general honeycomb material, which also mainly includes the elastic phase, stress plateau phase and densification phase. This is because the negative Poisson's ratio structure is a special kind of honeycomb multi-cell structure by nature, so it obeys the general change law of the material, but there are some differences between the two. In the elastic stage, the stress–strain curve of general honeycomb material tends to flatten out with increasing strain, and the stiffness of the structure gradually becomes smaller, while the slope of the curve of the negative Poisson's ratio structure gradually increases with increasing strain, so the stiffness of the structure also gradually increases, the difference between the two is mainly due to the different deformation of the microscopic cells and cell walls of the negative Poisson's ratio structure in the one-way compression process. In the stress plateau stage, the negative Poisson's ratio material can show higher plateau stress compared with the honeycomb material of the same basis material, which is mainly because in the stress plateau stage, the angle between the cell walls of the negative Poisson's ratio structure becomes smaller, which makes it more difficult for the cell walls to bend, thus increasing the failure load of the structure, and the negative Poisson's ratio structure shows higher plateau stress. In the stress plateau region, due to the special shape of the microtopology of the negative Poisson's ratio structure, the negative Poisson's ratio structure has a higher plateau stress compared with the traditional multi-cell structure, and its one-way compressive stress–strain curve can be further subdivided into two different stages, the stress plateau region and the stress plateau enhancement region. Unlike the conventional honeycomb structure, in the stress plateau enhancement region, the stress of the negative Poisson's ratio structure will continue to become larger as the strain increases, which will also have a greater impact on the energy absorption of the negative Poisson's ratio structure.

According to the relevant definitions of mechanics, it can be seen that the energy absorbed by the honeycomb structure per unit volume during the external force impact process in the stress–strain curve is the area enclosed by the stress–strain curve and the coordinate axis. The larger the area, the more energy the structure absorbs under the external force impact, and vice versa. In the linear elastic phase, the negative Poisson's ratio structure undergoes elastic bending deformation, and the stress of the structure gradually rises, and the negative Poisson's ratio structure absorbs the work done by the external force by generating deformation. After entering the stress plateau stage, the negative Poisson's ratio structure firstly enters the stress plateau zone, and as the stress level of the cell wall reaches the stress limit state, the cell wall gradually undergoes plastic collapse, and the plateau stress in this zone tends to a stable level. In the process of cell wall collapse, generally the first layer of cells collapses first, the gap between the layers is gradually filled, and the stress of the structure gradually decreases, then the second layer of cells collapses again, the stress rises again, and when the gap is filled, the stress gradually decreases, and the process is repeated until all cells in the structure collapse. Therefore, in the stress–strain diagram, the stress plateau area actually consists of several peaks, and because the size and distribution of the peaks are basically the same, the stress in this stage remains constant during the analysis. In the stress plateau interval, because the stress level has been maintained at a high level and the negative Poisson's ratio structure is in this interval for a longer period of time, the negative Poisson's ratio structure absorbs more energy in the stress plateau area, which has an important influence on the total energy absorption of the structure. The stress plateau enhancement zone of the negative Poisson's ratio structure is located after the stress plateau zone and before the densification stage. Due to the special microtopology of the negative Poisson's ratio structure, the material produces compression-shrinkage phenomenon, and the cells in different positions of the structure gradually approach the center, and when the strain in the stress plateau area exceeds a certain range and before the densification stage, the center of the structure forms a non-uniform multi-cell structure again, and the relative density of this part of the structure gradually becomes larger as the strain increases, which is shown in the stress–strain diagram The relative density of this part of the structure gradually increases with the increase of strain, which is shown in the stress–strain diagram as the platform stress becomes higher again. Since the average stress level in the stress plateau enhancement zone is higher than that in the stress plateau zone and has a certain action time in the process of external impact, the energy absorbed in this zone is also more. Under the action of external impact, with the increasing strain, the gap in the negative Poisson's ratio structure becomes smaller and smaller, the cell walls come into contact with each other, the stress of the structure rises rapidly, the relative density approaches the density of the basis material, and the structure enters the densification stage. Although the stress level of the negative Poisson's ratio structure is high in this stage, the energy absorbed in this stage is still low because the duration of this stage is short and the strain generated by the structure is small. The total energy absorbed by the structure at different stages of the process is the maximum energy that can be absorbed by the negative Poisson's ratio structure.

2.4.2 Energy Absorption Mechanism of Negative Poisson's Ratio Structure

In the process of external impact, the area enclosed by the stress–strain curve and the coordinate axis is the energy absorbed per unit volume of the negative Poisson's ratio structure, so the relationship between the energy absorbed by the structure and the parameters of the negative Poisson's ratio structure under the external force can be found by the stress–strain curve of unidirectional compression. Although the actual impact process of the negative Poisson's ratio structure subjected to external force is somewhat different from the quasi-static axial compression of the structure, the relationship between the energy absorption of the structure and the parameters under unidirectional axial compression can be obtained to guide the analysis of the energy absorption characteristics of the structure.

As shown in Fig. 2.17, the energy absorbed by a negative Poisson's ratio structure under external forces, i.e., the area between the stress–strain curve and the coordinate axis, can be expressed as

$$W = \int_{0}^{\varepsilon_S} \sigma(\varepsilon) d\varepsilon \tag{2.79}$$

where W is the energy absorbed by the structure and ε_S is the strain of the structure when the negative Poisson's ratio structure is about to enter the densification stage.

According to the analysis in Sect. 2.4.1, the negative Poisson's ratio structure absorbs less energy during the densification stage, so this energy is neglected in Eq. (2.79) to obtain the energy absorbed by the structure. In the stress–strain curve, the equivalent elastic modulus $\overline{E}$ is related to the equivalent force σ and equivalent strain $\overline{\varepsilon}$ of the negative Poisson's ratio structure as follows

$$\overline{\sigma} = \overline{E}\overline{\varepsilon} \tag{2.80}$$

In the stress–strain curve, the slope of the curve is the equivalent elastic modulus of the structure. $\overline{E}_A$ is the average elastic modulus of the negative Poisson's ratio structure, and the solution process assumes that the curve is linear within the elastic phase, then the average elastic modulus of the negative Poisson's ratio structure within the elastic phase $\overline{E}_{A1}$ can be expressed as

$$\overline{E}_{A1} = \frac{\overline{E}_0 + \overline{E}_{A2}}{2} \tag{2.81}$$

where $\overline{E}_0$ is the initial elastic modulus of the negative Poisson's ratio structure and $\overline{E}_{A2}$ is the average elastic modulus of the structure in the stress plateau region. The area enclosed by the stress–strain curve and the coordinate axis within the online elastic phase of the negative Poisson's ratio structure can be expressed as

$$W_1 = \frac{1}{2}\sigma_2\varepsilon_2 \tag{2.82}$$

where, σ_2 is the stress magnitude of the structure in the stress plateau region and ε_2 is the strain of the structure at the end of the linear elastic phase. Combining Eqs. (2.80), (2.81) and (2.82), the energy absorbed by the negative Poisson's ratio structure during the linear elastic phase can be found as

$$W_1 = \frac{\sigma_2^2}{\overline{E}_0 + \overline{E}_{A2}} \tag{2.83}$$

The stress plateau phase of the negative Poisson's ratio structure is divided into the stress plateau zone and the stress plateau enhancement zone, so the energy absorbed by the negative Poisson's ratio structure in this phase is also divided into the energy absorbed in the two zones together. In the stress plateau zone, the stress of the structure remains basically constant, so the absorbed energy in this zone can be expressed as

$$W_2 = \sigma_2(\varepsilon_3 - \varepsilon_2) \tag{2.84}$$

where, ε_3 is the strain of the negative Poisson's ratio structure at the end of the stress plateau zone. Assuming that $\overline{E}_{A3}$ is the average modulus of elasticity of the negative Poisson's ratio structure in the stress plateau zone, combined with Eq. (2.80), we find that

$$W_2 = \frac{\sigma_3\sigma_2}{\overline{E}_3} - \frac{2\sigma_2^2}{\overline{E}_0 + \overline{E}_{A2}} \tag{2.85}$$

In the stress plateau enhancement zone, the stress in the negative Poisson's ratio structure gradually increases with increasing strain. Using the approximation method in the online elastic phase of the structure, assuming that the structure is linear in the stress plateau enhancement region, the energy absorbed by the negative Poisson's ratio structure in this region can be expressed as

$$W_3 = \frac{1}{2}(\sigma_4+\sigma_3)(\varepsilon_4 - \varepsilon_3) \tag{2.86}$$

where σ_4 and ε_4 are the stresses and strains of the negative Poisson's ratio structure at the end of the stress plateau enhancement zone, combined with Eq. (2.80), the energy absorbed by the stress plateau enhancement zone can be expressed as Eq. (2.87). It should be noted here that, for the general multi-cell structure, the cells are gradually dispersed during the process of axial compression of the structure by external forces, resulting in the compression-expansion phenomenon, and the contact between the cell walls of the cells occurs at the end of the stress plateau zone, the stress rises rapidly, and the relative density approaches that of the basis material, and the duration of the structure in the stress plateau enhancement zone is very short, and the multi-cell structure rapidly enters the densification stage. In contrast, the negative Poisson's ratio structure, due to the process of axial compression, the cells gradually close to the center, resulting in compression-compression phenomenon, the structure in the stress plateau enhancement zone lasts longer, and the energy absorbed in this phase is more.

$$W_3 = \frac{(\sigma_4 - \sigma_3)(\sigma_4\overline{E}_3 - \sigma_3\overline{E}_4)}{2\overline{E}_3\overline{E}_4} \tag{2.87}$$

Although the stress in the densification phase of the negative Poisson's ratio structure is larger, but because the duration is very short, the energy absorbed by the structure in this phase is negligible, so the total energy absorbed by the negative Poisson's ratio structure in the process of unidirectional axial compression can be obtained by summing the energy absorbed by the structure in different phases.

By analyzing the stress–strain curves of the negative Poisson's ratio structure at different stages, the energy absorbed by the general honeycomb structure and the negative Poisson's ratio structure is investigated. The negative Poisson's ratio structure will significantly enhance the energy absorption capacity of the structure because of the increased stress plateau enhancement zone during the external force action, which is of great significance for the negative Poisson's ratio structure to be widely used in the design of automobiles with extremely limited design space. In the design process of the negative Poisson's ratio structure, in order to maximize the energy absorption capacity of the component, the structure should have a high stress level, and the stress–strain curve of the structure should have a long stress plateau area and stress plateau enhancement area.

The relative density of the negative Poisson's ratio structure has a large influence on the energy absorption of the structure. When the relative density of the structure is low, due to the larger gap between cells, the deformable space of the structure is larger, so the range of the stress plateau area is larger, and it is relatively difficult to enter the densification stage, and the stress plateau stage lasts longer, at the same time, when the relative density of the structure is low, it is also easy to produce a larger stress level during the deformation process, so the energy absorbed per unit structure is more. However, the lower relative density of the negative Poisson's ratio structure will reduce the stiffness and strength of the structure, and the application requires a larger space, so it is necessary to balance the properties of the negative Poisson's ratio structure. The energy absorption capacity of the negative Poisson's ratio structure is not only related to its own material and structure, but also related research shows that when the impact velocity of the negative Poisson's ratio structure increases, the inertia effect of the structure will be more obvious and the equivalent platform stress will be larger, thus the capacity of the structure to absorb during the impact will also be enhanced. Therefore, it is clear from the above analysis that the design of negative Poisson's ratio structure should not only consider the energy absorption, stiffness and cost of the structure, but also need to have a detailed understanding of the main range of impact velocity in the actual process, so as to maximize the advantages of the negative Poisson's ratio structure and improve the performance of the structure.

2.5 Summary of This Chapter

In this chapter, the negative Poisson's ratio structure is regarded as a continuous medium, and the macroscopic performance of the structure is described by the equivalent performance method, and the macroscopic performance of the structure is studied from the shape and organization form of the microstructure of the negative Poisson's ratio structure. By studying and analyzing the microstructure of the double-arrow negative Poisson's ratio structure, the microstructure shape parameters of the negative Poisson's ratio structure and the equivalent elastic model and the mechanical model of the equivalent Poisson's ratio are established, which is of great significance for mastering the relationship between the macro performance and microstructure parameters of the structure. By comparing and analyzing the negative Poisson's ratio structure models with different microstructure shape parameters, the static characteristics of the negative Poisson's ratio structure are studied, which provides a theoretical basis for better understanding the performance of the negative Poisson's ratio structure. The in-plane compressive mechanical properties of negative Poisson's ratio materials and traditional honeycomb materials were compared and analyzed, and the energy absorption characteristics and mechanism of negative Poisson's ratio materials were studied. The analysis of the mechanical properties and energy absorption mechanism of negative Poisson's ratio materials in this chapter is a necessary prerequisite for the subsequent simulation analysis, application and optimization of negative Poisson's ratio materials.

References

1. Dassault Systèmes, France. Abaqus Keyword Reference Guide [M]. Nanjing Southeast University Press 201509.504.
2. Guo Guangping, Ding Chuanfu, Beijing Institute of Aeronautical Materials, China Aerospace Development.Mechanical properties testing of aviation materials [M]. Mechanical Industry Press 2017 12. 414.
3. Meng Junhui, Liu Qingyang, Jin Zehua. Negative Poisson 's ratio metamaterial design based on complex constrained topology optimization [J]. Journal of Beijing Institute of Technology. 2023,43(08): 852–862.https://doi.org/10.15918/j.tbit1001-0645.2022.203.
4. Tang Chang. Tolerance design method for metamaterials with negative Poisson's ratio [D]. Dalian University of Technology. 2020. https://doi.org/10.26991/d.cnki.gdllu.2020.001387.
5. Research on mechanical properties of negative Poisson's ratio structure and its application in crashworthiness optimization design of commercial vehicles [D]. Nanjing University of Science and Technology. 2018.
6. Wang Qiang.Research on multi-condition crashworthiness of gradient negative Poisson 's ratio structure based on automobile energy absorbing box [D]. Jilin University. 2022. https://doi.org/10.27162/d.cnki.gjlin.2022.000919.
7. Majie. Design and mechanical properties of a new negative Poisson's ratio material structure [D]. Northeast University. 2018. https://doi.org/10.27007/d.cnki.gdbeu.2018.000246.

Chapter 3
Design Methods for Negative Poisson's Ratio Structures

3.1 Introduction

The microstructure of structures like the Poisson ratio and honeycomb structures plays a significant role in influencing their macroscopic performance. Researchers have been striving to establish a relationship between the microstructure of materials and their macroscopic properties to guide the design and optimization of material microstructures according to the desired performance in material applications. Gibson and others conducted research on individual microstructural units, developing structural models for discrete microstructural units and obtaining fundamental mechanical parameters for honeycomb materials. Hohe and others analyzed the elastic response of triangular, quadrilateral, and hexagonal honeycomb material structures. Grenestedt established a mechanical analysis model for the periodic regular honeycomb material's elastic properties. Gu and others studied the relationship between the microstructure of honeycomb structures and the material's thermal dissipation coefficient. In addition to the static performance of honeycomb materials, some scholars have also conducted research on the dynamic mechanical properties of honeycomb materials and their relationship with material microstructures. Liu Ying and others studied the influence of the micro-topological structure of honeycomb materials on their dynamic impact performance under in-plane impact loads. Zhang Xinchun and his team investigated the effects of cell expansion angles on the energy absorption and in-plane impact deformation of concave hexagonal honeycomb materials. These studies have greatly enhanced our understanding of the relationship between microstructural parameters and macroscopic performance in honeycomb structures, effectively guiding the design of microstructures for structures like negative Poisson ratio materials. However, due to the complexity of honeycomb structure performance, the uncertainty of the connection between microstructures, and the significant differences in the microstructural performance of various honeycomb materials, it is challenging to directly apply the established mechanical models to the design of microstructures for honeycomb structures. This makes it difficult to obtain

G. Zhou et al., *Design, Manufacturing, and Application of Negative Poisson's Ratio Structure*, https://doi.org/10.1007/978-981-96-1937-5_3

the optimal structure that meets performance requirements. Therefore, there is a need to find a more effective method for the design and optimization of microstructures in structures like negative Poisson ratio materials and other honeycomb structures. In macrostructure design, the application of various optimization methods to improve structural design for better performance is well-established. There are numerous optimization methods available with high precision and efficiency, allowing for the rapid identification of structures that meet design requirements. However, there are relatively few methods currently available for optimizing material microstructure parameters to achieve materials with superior performance. To obtain negative Poisson's ratio materials with better mechanical properties while considering the periodic regularity and structural characteristics of the double arrowhead-shaped negative Poisson's ratio microstructure, this paper introduces, for the first time, a parametric model for the symmetric microstructure of negative Poisson's ratio materials. It combines the Design of Experiments (DOE) method and the response surface methodology to establish a response surface model for the double arrowhead-shaped negative Poisson's ratio structure. Subsequently, optimization algorithms are used to optimize shape parameters and thickness of the microstructure, thereby obtaining a negative Poisson's ratio structure that meets specific design requirements. While the connectivity of microstructures in honeycomb materials and the number of adjacent units also have significant effects on the mechanical properties of materials, the geometric design parameters of microstructures in honeycomb materials, such as the length, thickness of cell walls, and the angles of connection between cell walls, have a more direct impact on the mechanical performance of honeycomb structures. The double arrowhead-shaped negative Poisson's ratio structure studied in this paper possesses a microstructure with periodic regularity, and the interconnections between microstructures are essentially determined. By varying the microstructural parameters of the double arrowhead-shaped negative Poisson's ratio structure, it is possible to obtain a negative Poisson's ratio structure that meets specific design requirements, such as having maximum stiffness, maximum negative Poisson's ratio, or minimum stress. Therefore, optimizing microstructural parameters in a rational manner during the design process is of significant importance for achieving improved performance in negative Poisson's ratio structures.

3.2 Single-Objective Optimal Design of Negative Poisson's Ratio Structure

In practical engineering applications, materials are required to meet strict performance criteria across various aspects, including stiffness, stress, mass, and durability. Negative Poisson's ratio structures, while possessing unique mechanical properties, also have certain drawbacks. Traditional negative Poisson's ratio materials often exhibit lower stiffness and strength due to their relatively large deformation spaces within the material, making them susceptible to stress concentration under external loads. Additionally, there is a trade-off between the special properties of negative Poisson's ratio materials and other material properties. Materials exhibiting negative Poisson's ratio characteristics are generally porous, with sufficient internal space to allow for cell wall rotation, bending, buckling, and hinging. However, increasing the material's stiffness and strength can often lead to the loss of negative Poisson's ratio properties. In summary, developing negative Poisson's ratio structures that possess both negative Poisson's ratio characteristics and reasonable stiffness, strength, and weight is a crucial topic in this research field. This section focuses on optimizing the microstructural parameters of the double arrowhead-shaped negative Poisson's ratio structure to meet different design requirements and improve the macroscopic performance of negative Poisson's ratio structures.

3.2.1 Study of Single-Objective Optimization Method and Process for Negative Poisson's Ratio Structure

In this chapter, based on the shape and arrangement characteristics of the microunits in the double arrowhead-shaped negative Poisson's ratio structure, we propose a microstructure optimization design method for negative Poisson's ratio structures based on macroscopic performance. The optimization process is illustrated in Fig. 3.1.

In the optimization process of the double-arrow type negative Poisson's ratio structure, the parametric model of the negative Poisson's ratio structure is firstly established and the accuracy of the analysis results of the parametric model is verified, followed by the definition of different design objectives, variables and constraints, etc. in the optimization process to establish the optimization model. Considering the complexity of the microstructure of the double-arrow type negative Poisson's ratio structure, a large negative Poisson's ratio structure will lead to a great amount of computation, making the analysis and optimization process less efficient or even impossible. After verifying the response surface model, the parameters of the negative Poisson's ratio microstructure are optimized using genetic algorithms and other optimization algorithms to obtain a better performance of the double-arrow negative Poisson's ratio structure. The parameters of the negative Poisson's ratio structure are optimized using genetic algorithms and other optimization algorithms to obtain a double-arrow negative Poisson's ratio structure with better performance. The parametric model of the negative Poisson's ratio structure has been discussed in detail and the following is a brief description of the experimental design method, response surface method, and optimization algorithm used in the optimization of the double-arrow negative Poisson's ratio structure.

DOE (Design of Experiment) plays an important role in the process of constructing approximate models. The design of experiment focuses on the rational arrangement of the experiment, the selection of a limited number of samples in the entire design area, and the acquisition of desirable test results and scientific conclusions with

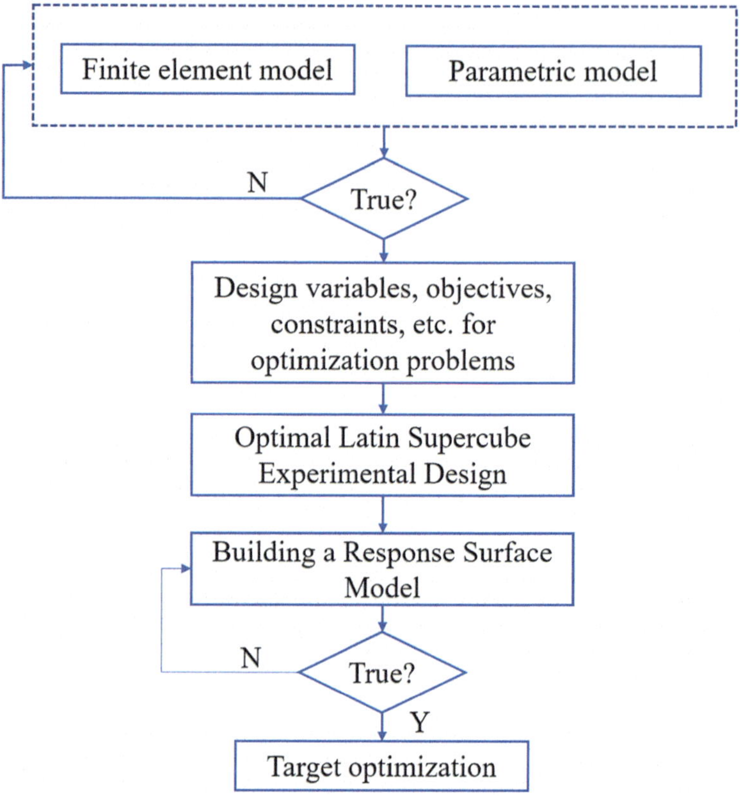

Fig. 3.1 Optimization process of negative Poisson's ratio structure based on intrinsic properties

a smaller test size (number of tests), shorter test period, and lower test cost. It is mainly used to study the effect of each design parameter on the response and use it to make appropriate decisions [1]. The appropriateness of sample point selection affects the accuracy of the approximate model to a certain extent, so the selection of sample points in the experimental design is very important in the construction of the approximate model. Commonly used experimental design methods are: partial factorial experimental design, full factorial experimental design, central composite design, Latin square experimental design and optimal Latin square experimental design, etc. [2].

The Latin square experimental design samples uniformly in the design space of each factor and then randomly combines the levels, which has the advantages of effective space filling ability and better fitting nonlinear response compared with the traditional experimental design method, but with the increase of the number of test levels, there is a disadvantage of uneven distribution of test points when facing the high-dimensional design space due to the random combination property. The optimal Latin square experimental design improves on the Latin square experimental design by optimizing the order of occurrence of each level in each column of the design matrix [3], so that the factor levels of each sample point in the matrix are more evenly distributed, making the fitting of factors and responses more accurate and realistic [4]. The general optimization criterion in the process of optimizing the design matrix is as follows.

$$Y(x) = \sum_{i=1}^{n} \alpha_i f_i(x) + Z(x) \tag{3.1}$$

In the above equation, $Y(x)$ is the response; $Z(x)$ is a Gaussian distribution function with mean 0. The covariance function of the l-dimensional input s with t can be expressed as

$$R(s, t) = \sigma^2 \exp\left(-\theta \sum_{j=1}^{l} \left|s_j - t_j\right|^q\right), 0 < q \leq 2 \tag{3.2}$$

The performance of $Z(x)$ in the above equation depends on the parameters θ and q, and the optimization criterion in the sample point design process is to minimize $-lg|R|$. Figure 3.2 shows the comparison between a full factorial design with two factors (X1, X2) and nine sample points, Latin square sampling and optimal Latin square sampling.

Approximate modeling is a method of approximating a set of design and response variables through mathematical modeling based on sample data or design-of-experiment sampling techniques, thus avoiding intensive simulation calculations and reducing iteration time. Response Surface Method (RSM) is an important approximate modeling method that simulates a real limit state surface by fitting a response surface through a series of deterministic "experiments" [5]. Assuming that the dependent variable Y and the independent variable X satisfy $Y = F(X)$, where X is an n-dimensional independent design variable, $X = [X_1, X_2, \ldots, X_n]$, by using the response surface model approximation, the above functional relationship is: $Y = F(X)' + \varepsilon$, $F(X)'$ is the response function on the variable X, ε is the random

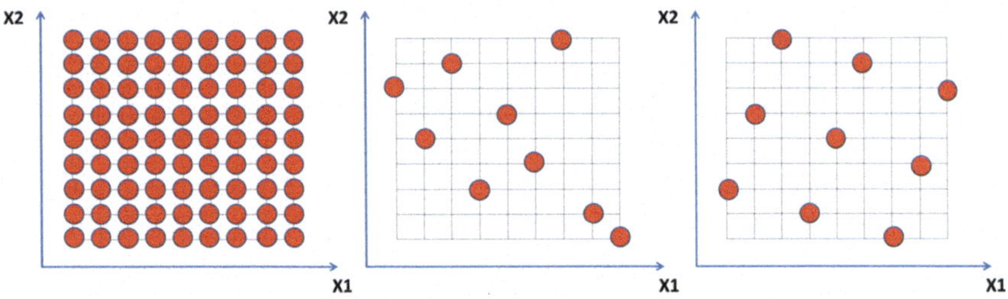

Fig. 3.2 Comparison plots of full factorial design, Latin square sampling and optimal Latin square sampling

error between the approximation and the actual value, ε obeys the $\left(0, \varepsilon^2\right)$ standard normal distribution.

The moving least squares response surface method is an improvement on the traditional response surface method. Although the regression calculation method is still used, the coefficients of each item in the approximate model construction process will change with the change of sampling points, so that it is less prone to distortion and has better fitting accuracy when dealing with larger discrete data [6]. The response surface model is generally of order 1–4, and the higher the order, the better the ability to fit the nonlinear model. The minimum number of sample points for constructing a response surface model is determined by the number of input variables selected and the order of the model, which is commonly used as a second-order response surface model, which requires (n + 1) × (n + 1)/2 exact analyses and has high accuracy. The second-order response surface model can be expressed as follows.

$$F(x) = a_0 + \sum_{i=1}^{N} b_i x_i + \sum_{ij(i<j)} c_{ij} x_i x_j + \sum_{i=1}^{N} d_i x_i^2 \tag{3.3}$$

The values of *a*, *b*, c, and d are determined by using the moving least squares method, i.e., by finding the minimum value for the weighted residual function. The process of finding the weight function of the sampling points is required to ensure the continuity and localization of the model approximation. The selection of the weight function plays an important role in the construction of the moving least squares approximation model [7]. In general, the weight function needs to satisfy semi-normality, normalization, tight branching, decreasing and normalization. The commonly used weight functions mainly include Gaussian weight function, cubic spline function, quadratic spline function and exponential function, and Gaussian weight function is mainly used in engineering. Suppose the radius of the support domain is d_{max}, $s = d/d_{max}$, then the Gaussian weight function can be expressed as follows.

$$w(s) = \begin{cases} \frac{\exp(-s^2\beta^2)-\exp(-\beta^2)}{1-\exp(-\beta^2)} & 0 < s < 1 \\ 0 & s > 1 \end{cases} \tag{3.4}$$

For the optimization problems in engineering, when the response function value oscillates, it is difficult to obtain the response surface function with high accuracy because the traditional polynomial response surface method is based on the global approximation method and has high nonlinearity in space [8]. The moving minimum response surface method, however, can still obtain a high accuracy response surface model when dealing with response oscillations because the coefficients of each item in the construction process can be varied. Of course, the drawback is that the approximation process is computationally intensive because of the large number of calculations for each prediction point, but with the rapid upgrading of computer hardware and software, the problem has been better solved. The F-test and the R^2 test are generally used in the process of approximating the model to detect the accuracy of the response surface construction, where the R^2 test is mainly to see how close R^2 is to 1. The closer they are, the higher the accuracy of the approximation model. Since it is easy to obtain the results of R^2 test, the R^2 test is chosen to judge the accuracy of the approximate model.

Genetic algorithms are an efficient optimization algorithm that is similar to the principles of evolution and natural selection observed in the biological world. The concept of genetic algorithms was initially proposed by Professor Holland at the University of Michigan in 1975 [9]. These algorithms work by rapidly searching for optimal solutions through processes such as selection, crossover, and mutation of genes. In biology, fitness is commonly used to measure the adaptability of a species to its environment. In the context of genetic algorithms, fitness is employed to evaluate how well each individual in a population approaches the optimal or best quality during the optimization process. If an individual has a high fitness, it has a greater probability of passing its genes to the next generation in the optimization process. Conversely, individuals with low fitness have a lower probability of being passed on to the next generation. This concept of fitness and the genetic operations of selection, crossover, and mutation enable genetic algorithms to efficiently explore and converge towards optimal solutions in various optimization problems [10].

For the optimization of finding the minimum value of the response, the genetic algorithm fitness function can be expressed as follows.

$$F(\overline{X}) = \begin{cases} C_{\max} - f(\overline{X}) \\ 0 \end{cases} \tag{3.5}$$

where $f(\overline{X})$ is the objective function and C_{max} is a larger number of appropriate choices. The optimal conservation strategy and proportional selection can be used as selection operators in the optimization process to determine which individuals will be inherited into the next generation population. In the crossover process, arithmetic crossover can be used to achieve the crossover of two individuals, e.g., two individuals P_A^n and P_B^n can be crossed over to produce new individuals P_A^{n+1} and P_A^{n+1}, where δ is a random number between 0 and 1. Also for the genetic variation, the variation probability is usually set before the optimization of the problem. In order to ensure a fast and stable search for the optimal value, the variation probability should be chosen as a suitable value, which is set to 0.01 in this section.

$$\begin{cases} P_A^{n+1} = \delta P_B^n + (1-\delta) P_A^n \\ P_A^{n+1} = \delta P_A^n + (1-\delta) P_B^n \end{cases} \tag{3.6}$$

Compared to other optimization algorithms, genetic algorithms have distinct advantages. Firstly, genetic algorithms do not require problem-specific information or details directly related to the problem. They require minimal a priori information and are capable of generating general-purpose algorithms that can handle a wide range of optimization problems. Secondly, genetic algorithms, when they begin the search, start from an initial population, which covers a broad search space and is conducive to finding global optima. They possess a wide-ranging global search capability. Furthermore, during the optimization process, genetic algorithms can quickly eliminate values that deviate significantly from the optimal solution through selection, crossover, and mutation. They exhibit strong fault tolerance. However, it is essential to employ effective strategies during the optimization process to prevent the algorithm from getting stuck in local optima [11].

3.2.2 Parametric Modeling of Negative Poisson's Ratio Structure

In traditional structural finite element analysis and modeling, a unidirectional modeling approach is typically used. This approach involves gradually building a finite element analysis model based on predetermined structural parameters or a three-dimensional model. Subsequently, relevant analysis and optimization are performed based on the established model. This method has the advantage of high model accuracy and the ability to effectively reflect changes in structural performance. It works well for simple structural modifications and validations [12]. However, when dealing with larger models or the need to optimize structural parameters such as shape, the inability to automatically modify models generated from three-dimensional geometric data can make model modification and replacement very challenging. Even minor changes to the data often require the entire modeling and analysis process to be repeated, which is time-consuming and labor-intensive. This is not conducive to the efficiency and rapid analysis required for multi-scheme design in engineering. Traditional finite element modeling and structural optimization can become increasingly difficult in such scenarios [13].

To achieve rapid simulation, modeling, and optimization of complex structures while ensuring that the structural shape can be quickly constructed, modified, and replaced during analysis, the concept of simulation-driven design can be realized through parametric modelling [14]. In parametric modeling, by setting different parameters for various structural attributes such as shape and thickness, different finite element analysis models can be rapidly created by assigning different data to these parameters. Based on a fundamental parametric model, rapid solid modeling and subsequent analysis can be carried out. By utilizing parametric modeling, the speed of creating similar models can be significantly increased, facilitating the extension of analysis for different parameters of the same model. This approach enables rapid simulation, modeling, and optimization of complex structures, thereby reducing structural optimization design time and enhancing work efficiency [15].

In parametric design, critical structural parameters are typically chosen as design parameters for parametric modeling. The relationships between these parameters are established to define the shape and topology relationships of different parts within the structure. This enables the creation and modification of parametric structural models. Therefore, when a parameter of a component is modified, the shape parameters of other parts of the structure can also be automatically adjusted, significantly improving modeling efficiency. Parametric modeling is a key component of parametric optimization design, and its rationality and accuracy directly impact the complexity and quality of parametric optimization design [16]. Hence, selecting an appropriate parametric modeling strategy is of paramount importance. It not only enhances the efficiency and accuracy of parametric modeling but also facilitates the invocation of parametric optimization routines and structural optimization design. In the parametric modeling process, not only can the finite element mesh model of the structure be parameterized, but the subsequent analysis processes in which finite element software uses the mesh model for analysis can also be parameterized. This includes aspects such as analysis type, structural load application, material property definitions, constraints, and the extraction of analysis results.

During the structural optimization design process, it is essential to analyze various design parameters and select appropriate parameters for constructing an efficient parametric model. To enhance the efficiency and accuracy of structural optimization design, parametric models should meet the following three requirements:

(1) The model is driven by design parameters and the FEA model of the structure can be changed by changing the parameters.
(2) The ability to quickly obtain the changed design parameters from external databases or files, and the ability to output the analysis model and parameters for finite element calculations to the external world after generating the model.
(3) The parametric model program can run independently and has low dependence on other programs. The optimization process can be easily called by other optimization programs to generate new finite element models that can be used for analysis and design.

The double-arrow negative Poisson's ratio structure studied in this paper has a periodic regular microstructure, and the negative Poisson's ratio structure in practical application contains a large number of microscopic units, which takes a long time to model and has a large computational volume, and it is difficult to analyze and optimize the double-arrow negative Poisson's ratio structure effectively by traditional methods such as trial-and-error method. Therefore, in this paper, we propose a parametric model to change the parameters of the negative Poisson's ratio structure and establish the computational model quickly, so as to improve the modeling efficiency of the negative Poisson's ratio structure and shorten the optimization design time.

In order to avoid the need to re-model the negative Poisson's ratio structure for each parameter change during the research process, which causes huge repetitive workload and time wastage, this subsection realizes the whole process of analytical modeling work by writing a parametric program for the double-arrow type negative Poisson's ratio structure, and the modeling process only needs to change the design parameters in the program to realize the automatic modeling and loading analysis condition setting of the negative Poisson's ratio structure [17]. The microstructure of the double-arrow negative Poisson's ratio structure studied in this paper is geometrically symmetric, and its key parameters include the length of the long cell wall l_1, the angle between the long cell wall and the symmetry axis of the microstructure θ, the angle between the short cell wall and the symmetry axis of the microstructure ϕ, the thicknesses of the long and short cell walls T_1 and T_2, and the width of the cell wall a. Using the above key parameters of the microstructure of the negative Poisson's ratio structure, the finite element model of the 4×4 double-arrow negative Poisson's ratio structure is automatically established and the boundary conditions are automatically loaded by a self-programmed Matlab program. The whole parametric modeling process does not require human intervention, which greatly improves the speed of establishing the analytical model of negative Poisson's ratio structure, and the program has good independence and is compatible with other finite element analysis software and optimization programs.

The design concept and modeling process of the parameterized design program for the double arrow negative Poisson's ratio structure in this section are illustrated in Fig. 3.3. To begin, you input the structural modeling parameters in an interactive interface or a text document. Since the design of the negative Poisson's ratio structure imposes certain restrictions on parameter ranges and their interrelationships, you need to verify whether the input parameters are reasonable. If the parameters are not valid, the program will prompt you to re-enter them. Once the input conditions are met, the program calculates the values of non-deterministic shape parameters in the negative Poisson's ratio structure based on the relationships between microstructure parameters. With the input parameters and calculated shape parameters, the program first establishes the model of a single cell unit structure for the double arrow negative Poisson's ratio structure. Then, by replicating this unit structure in different dimensions, it constructs the finite element mesh model of the negative Poisson's ratio structure. Subsequently, it automatically applies the predefined material properties and boundary loads to the mesh model to ensure that the finite element model is ready for simulation analysis. Finally, the determined model is exported in the required format for a general finite element model, marking the completion of the parameterized modeling for the negative Poisson's ratio structure. The parameterized model of the 4×4 single cell double arrow negative Poisson's ratio structure established in this chapter is illustrated in Fig. 3.4.

To validate the accuracy of the parameterized model, this study conducted quasi-static compression tests on specimens of the double arrow negative Poisson's ratio structure. The test specimens were manufactured using a 3D printer, and ABS plastic was used as the material for the test specimens, consistent with the material settings in the parameterized model. In the experimental process, the compression speed of the test rig was set to 1 mm/min to simulate the quasi-static compression process. The experimental results were compared with the analytical results from the parameterized model, as shown in Table 3.1. From the table, it can be observed that there is a certain level of error between the analytical results of the parameterized model and the results of the quasi-static compression test. The horizontal displacement error between the two is 12.56%, while the vertical error is 11.45%. For comparative

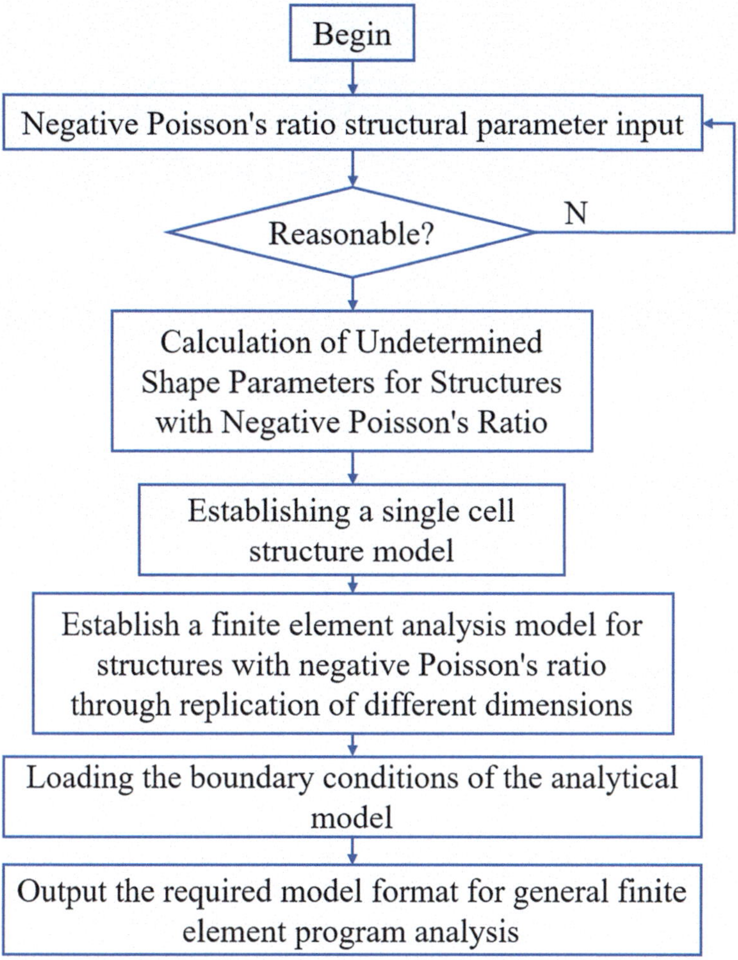

Fig. 3.3 Parametric modeling process of double-arrow type negative Poisson's ratio structure

analysis of the accuracy of the parameterized model, a 4 × 4 hexahedral element finite model of the double arrow negative Poisson's ratio structure was also created and analyzed, with the results presented in Table 4.1. Comparative analysis reveals that the hexahedral element finite element model's analytical results exhibit a lower error when compared to the results of the quasi-static compression test. Specifically, the horizontal error is 8.65%, and the vertical error is 9.99%. Therefore, the hexahedral element finite element model provides results that are closer to the experimental results in both the horizontal and vertical directions when compared to the parameterized model. The main sources of error in the parameterized model are two fold. First, the parameterized modeling process involved some simplifications at the

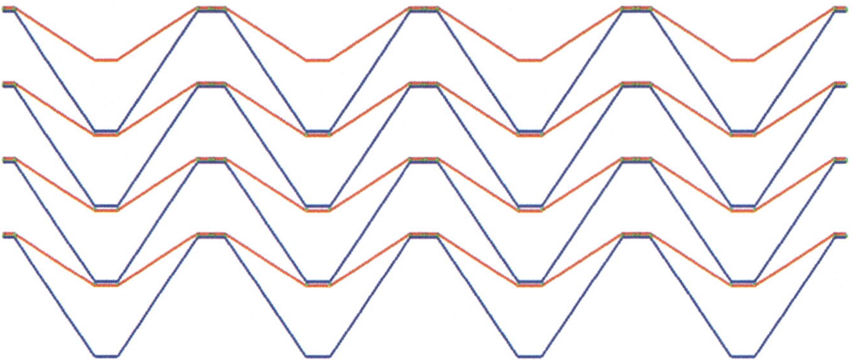

Fig. 3.4 Parametric model of double-arrow type negative Poisson's ratio structure

connection points of the microstructure cell walls of the negative Poisson's ratio structure, leading to increased errors. Second, the beam model used for the parameterized model has lower simulation accuracy compared to the hexahedral element model. As a result of these two reasons, the parameterized model exhibits lower accuracy when compared to the hexahedral element finite element model. Despite the slightly lower accuracy of the parameterized model for the double arrow negative Poisson's ratio structure compared to a general finite element model, the parameterized model still offers a high level of precision that meets the analysis requirements for this structure. Additionally, the parameterized model simplifies the modeling process, increases efficiency, and significantly reduces the analysis and optimization time for the negative Poisson's ratio structure. Therefore, this study continues to employ the parameterized model for the modeling and optimization of the double arrow negative Poisson's ratio structure.

Due to the ability of the parameterized model for the double arrow negative Poisson's ratio structure to rapidly establish new finite element models and output them in the required format for general finite element models based on parameter changes, it can be seamlessly integrated with various finite element software and optimization programs. This integration allows for the analysis of the macroscopic properties of the negative Poisson's ratio structure and the optimization of microstructure parameters, further enhancing the performance of the double arrow negative Poisson's ratio structure. In the next section, we will address the diverse performance requirements of the negative Poisson's ratio structure by combining the

Table 3.1 Comparison of analytical results of parametric models for double-arrow type negative Poisson's ratio structures

	Quasi-static compression experiments	Parametric model	Hexahedral unit finite element model
Level/mm	13.06	11.42	11.93
Vertical/mm	8.91	7.89	8.02

parameterized model and optimization programs to optimize the microstructure for different objectives.

3.2.3 Stiffness Maximization Design of Negative Poisson's Ratio Structure

The ability of a structure or material to resist deformation when subjected to external forces is referred to as its stiffness. In order to maintain the shape of a structure, natural structures in the environment must possess a certain level of stiffness. This is especially crucial in engineering designs like automobiles. Inadequate stiffness in a structure can lead to instability, vibration, stress concentration, fatigue, and crack formation [18], all of which can affect the structure's safety and functionality. Material stiffness is primarily measured using the elastic modulus, while structural stiffness is determined by the ratio of the loads the structure can bear to external forces, meaning the force required for a unit displacement of the structure [19]. Structural stiffness is closely related to factors such as the structural geometry, material properties, boundary conditions, and direction of applied loads. Therefore, in engineering structural design, it is essential to ensure that the structure possesses the required level of stiffness according to design specifications. In the context of this study, the double arrow negative Poisson's ratio structure is prone to insufficient stiffness due to the limited material filling within. Therefore, when applying this structure in relevant designs, it is crucial to ensure that the structure has sufficient stiffness to resist external deformations.

The stiffness of the double arrow negative Poisson's ratio structure is primarily influenced by the basis material of the structure, the geometric parameters of the arrow-shaped unit cells, and the number and arrangement of these units in the structure. Since the material used in engineering design is typically predetermined, this section primarily focuses on optimizing the stiffness of the structure under the conditions where the basis material and composition of the negative Poisson's ratio structure are established. The goal is to maximize the stiffness of the structure. In this section, the parameterized model constructed in Sect. 4.3 is used as the computational model. The key parameters in the parameterized model include the length l_1 of the long unit cell wall, the angles θ and ϕ between the long unit cell wall and the structure's axis of symmetry, the thickness T_1 and T_2 of the long and short unit cell walls, and the width a of the cell wall. Since the width of the cell wall (a) has a significant impact on the shape, weight, and structural performance of the negative Poisson's ratio structure, this study selects l_1, θ, ϕ, T_1, T_2, and a as the optimization design variables. The initial values and ranges of variation for each variable are provided in Table 4.2. In the design of the negative Poisson's ratio structure, to ensure that the structure exhibits the negative Poisson's ratio effect, it is essential that angles w and e be less than $\pi/2$, and angle a should be less than angles (Table 3.2).

Table 3.2 Stiffness maximization optimization design variables

Design variable	Initial	Minimum	Maximum
l_l/mm	18.50	5	30
θ/rad	0.34	0	1.57
φ/rad	0.56	0	1.57
a/mm	8.50	1	15
T_1/mm	2.50	0.1	4
T_2/mm	0.80	0.1	4

In the structural optimization process, it is essential to meet design requirements, ensuring the overall mass of the negative Poisson's ratio structure is optimized to achieve lightweight design and reduce factors such as fuel consumption and emissions in automotive applications. However, the stress levels in the structure must remain within permissible design limits to ensure durability. Since stiffness represents the force required for a unit displacement in the structure, the optimization problem of maximizing stiffness in this process can be translated into minimizing the displacement of the structure under a given load. In this section, the optimization problem of maximizing stiffness comes with constraints on the mass (m) and maximum stress (S) of the negative Poisson's ratio structure. Specifically, the mass (m) should not exceed 20 g, and the maximum stress (S) should not exceed 170 MPa. The mathematical model for the stiffness maximization optimization problem of the negative Poisson's ratio structure can be described as follows:

$$\begin{gathered}\text{Minimize d}\\ \text{Subjected to :}\\ \mathrm{m} < 20\,\mathrm{g},\ \mathrm{S} \leq 170\,\mathrm{Mpa}\\ 0 < \theta < \mathrm{j} < 1.57\\ 1 < \mathrm{a} < 15, 5 < \mathrm{l_l} < 30\\ 0.1 < \mathrm{T_1} < 4, 0.1 < \mathrm{T_2} < 4\end{gathered} \tag{3.7}$$

The initial shape parameters, initial mass, and initial displacements of the negative Poisson's ratio structure are listed in Table 3.3. During the optimization process, the accuracy of the approximate model, measured by the R-squared (R2) value, was found to be 0.984, indicating that the accuracy of the approximate model meets the requirements for the optimization design. After 42 iterations, the optimization results for the negative Poisson's ratio structure were obtained, and the comparison between the initial values and the optimization results for the structure's parameters is presented in Table 3.3.

From the comparison of the optimization results in the table above, it can be observed that under the same external force, the displacement of the double-arrowhead negative Poisson's ratio structure decreased from 7.89 to 5.94 mm, representing a reduction of 24.7%. Therefore, through the optimization of microstructure

Table 3.3 Comparison of optimization results of negative Poisson's ratio structure based on maximization of stiffness

Parameter	Initial	Optimization
l_l/mm	18.50	22.45
θ/rad	0.34	0.41
φ/rad	0.56	0.68
a/mm	8.50	7.27
T_1/mm	2.50	2.24
T_2/mm	0.80	0.64
d/mm	7.89	5.94
m/g	17.63	19.94
S/MPa	157.63	149.36

parameters, the stiffness of the negative Poisson's ratio structure has been significantly improved. After optimization, the maximum stress of the negative Poisson's ratio structure decreased from 157.63 to 149.36 MPa, indicating a slight stress reduction. Although the mass of the structure increased from 17.63 to 19.94 g, it still satisfies the design constraints. In summary, by optimizing the stiffness of the double-arrowhead negative Poisson's ratio structure, the structure's resistance to external deformation has been enhanced while meeting the constraint conditions.

3.2.4 Optimization of Maximum Stress Minimization for Negative Poisson's Ratio Structures Design

In general, when structures are subjected to external forces, internal forces interact within the structure to resist external influences and restore the structure to its initial state from a deformed state. The stress within a structure when subjected to external forces depends not only on the magnitude and direction of the external forces but is also closely related to the composition of materials, shape, and dimensions of the structure [20]. Because of internal stresses, structures are susceptible to issues such as cracking and bending deformation when subjected to external forces. If the stress levels are too high, the structure may undergo plastic deformation or even fracture, resulting in damage. This significantly impacts the mechanical properties, durability, and overall performance of the structure [21]. Therefore, it is crucial to pay special attention to the maximum stress within the structure during the design process, as most problems related to high stress levels typically occur at the location of maximum stress when subjected to external forces.

The maximum stress of the double-arrowhead negative Poisson's ratio structure under external forces is influenced by the base material, the shape parameters of the arrowhead-shaped unit cell, and its composition, especially the connections between cell walls. In order to reduce the maximum stress in the double-arrowhead negative Poisson's ratio structure under external forces, this section focuses on minimizing the maximum stress when optimizing the structure. It is assumed that the base material of the negative Poisson's ratio structure has already been determined. Therefore, the reduction of the maximum stress in the structure is achieved by optimizing the shape parameters and thickness of the micro-topology of the negative Poisson's ratio structure. The selection of variables and the range of variable variations in the optimization process are consistent with the stiffness maximization optimization discussed in Sect. 3.4.2. To ensure that the stiffness of the double-arrowhead negative Poisson's ratio structure remains at an appropriate level and to achieve lightweight design, the mass and stiffness of the structure are used as constraint conditions during the optimization process. The mass is constrained not to exceed 20 g, and the stiffness is constrained so that the displacement generated under external forces does not exceed 9mm. Therefore, the mathematical model for the problem of minimizing maximum stress can be represented as follows:

$$\begin{gathered} \text{Minimize S} \\ \text{Subjected to :} \\ \mathrm{m} < 20\,\mathrm{g}, d \leq 9 \\ 0 < \theta < \varphi < 1.57 \\ 1 < \mathrm{a} < 15, 5 < \mathrm{l}_1 < 30 \\ 0.1 < \mathrm{T}_1 < 4, 0.1 < \mathrm{T}_2 < 4 \end{gathered} \tag{3.8}$$

The initial shape parameters of the negative Poisson's ratio structure during the optimization process were kept consistent with the optimization of maximizing the stiffness of the negative Poisson's ratio structure in Sect. 3.2.3. The optimization results were obtained after 49 optimization iterations, and the comparison of the initial values and optimization results of the negative Poisson's ratio structure is shown in Table 3.4.

From the optimization results mentioned above, it can be observed that the total mass of the double-arrowhead negative Poisson's ratio structure increased from 17.63 to 19.43 g. However, the maximum stress in the structure decreased from 157.63 to

Table 3.4 Comparison of optimization results for negative Poisson's ratio structures based on maximum stress minimization optimization

Parameter	Initial	Optimization
l_l/mm	18.50	19.4
θ/rad	0.34	0.28
φ/rad	0.56	0.51
a/mm	8.50	7.61
T_1/mm	2.50	2.98
T_2/mm	0.80	1.23
d/mm	7.89	7.24
m/g	17.63	19.43
S/MPa	157.63	132.85

132.85 MPa, and the displacement produced by the negative Poisson's ratio structure under external forces decreased from 7.89 to 7.24 mm. Therefore, it is evident that through the optimization of microstructural shape parameters and cell wall thickness of the negative Poisson's ratio structure, the structure has successfully reduced the maximum stress level while meeting the mass requirements of the design. Additionally, the stiffness of the structure has been effectively improved, thus meeting the optimization design requirements.

3.2.5 Optimization of Negative Poisson's Ratio Structural Mass Minimization Design

Lightweighting in the automotive industry has become a crucial research direction in vehicle design and manufacturing. In automotive design and manufacturing, the overall vehicle mass is directly linked to factors such as fuel efficiency, emissions, and costs. According to relevant studies, reducing the weight of a vehicle leads to improved fuel economy, reduced fuel consumption, and lower tailpipe emissions [22]. For competing models within the same class, lower vehicle weight not only results in improved fuel efficiency and reduced emissions, making it more environmentally friendly, but it also holds significant importance for automakers in terms of cost accounting due to the high production volumes of individual car models [23]. Moreover, a lighter vehicle of the same class is more competitive in the market. This is particularly relevant given the large production volumes in the automotive industry, strict national regulations, and consumer demands for lower fuel consumption. The goal of automotive lightweighting is to reduce vehicle and component mass as much as possible while still meeting strength, stiffness, and other performance requirements. This reduction in weight helps lower fuel consumption, emissions, and production costs. The double-arrowhead negative Poisson's ratio structure, as a promising material for the automotive industry, needs to be designed in a way that reduces its mass while ensuring it meets the necessary performance criteria. This will make it more competitive in the market, considering the high production volumes and the stringent requirements imposed by national regulations and consumer preferences for reduced fuel consumption.

The mass of the double-arrowhead negative Poisson's ratio structure is not only related to the base material and cell wall thickness but also closely tied to the shape parameters of its micro-topology. In this section, when optimizing the design of the double-arrowhead negative Poisson's ratio structure, the primary goal is to minimize the structure's mass while ensuring that the maximum stress in the structure remains below the maximum allowable value and that the stiffness requirements are met. The selection of variables and the range of variable variations during the structural optimization process are the same as discussed in Sect. 3.4.2. As constraint conditions during the optimization process, it is required that the displacement produced by the negative Poisson's ratio structure under external forces does not exceed 9mm, and

Table 3.5 Comparison of optimization results of negative Poisson's ratio structures based on mass minimization optimization

Parameter	Initial	Optimization
l_l /mm	18.50	16.86
θ/rad	0.34	0.29
φ/rad	0.56	0.54
a/mm	8.50	8.9
T_1/mm	2.50	2.37
T_2/mm	0.80	0.95
d/mm	7.89	8.46
m/g	17.63	14.79
S/MPa	157.63	165.77

the maximum stress should not exceed 170 MPa. Therefore, the mathematical model for the optimization design of minimizing the mass of the negative Poisson's ratio structure can be represented as follows:

$$\begin{gathered} \text{Minimize m} \\ \text{Subjected to :} \\ \mathrm{S} \leq 170\,\mathrm{MPa},\ \mathrm{d} \leq 9 \\ 0 < \theta < \varphi < 1.57 \\ 1 < \mathrm{a} < 15, 5 < \mathrm{l_l} < 30 \\ 0.1 < \mathrm{T_1} < 4, 0.1 < \mathrm{T_2} < 4 \end{gathered} \tag{3.9}$$

The initial shape parameters for the double-arrowhead negative Poisson's ratio structure are the same as the ones used for the initial parameters based on stiffness and stress optimization. After 37 optimization iterations, the optimization results have been obtained. Table 3.5 provides a comparison between the initial parameters and the optimized results for the double-arrowhead negative Poisson's ratio structure.

Analysis of the optimization results indicates that after optimization, the total mass of the double-arrowhead negative Poisson's ratio structure has decreased from 17.63 to 14.79 g. Simultaneously, the deformation of the structure increased from 7.89 to 8.46 mm, and the maximum stress in the structure increased from 157.63 to 165.77 MPa. Although the deformation and stress in the optimized structure are higher than in the initial design, they still meet the design requirements. In summary, through the optimization of the microstructural parameters of the double-arrowhead negative Poisson's ratio structure, the mass of the structure has been maximally reduced while ensuring that the structure's strength and stiffness meet the design requirements. This is significant for the application of negative Poisson's ratio structures in lightweight automotive design.

3.3 Multi-objective Optimization of Negative Poisson's Ratio Structures Design

In practical engineering applications, structural design often necessitates the simultaneous consideration of multiple distinct design objectives. These different design objectives may impose varying or even conflicting requirements on design parameters, thus requiring a balance and trade-off among these objectives. Therefore, this section focuses on the multi-objective optimization design of the microstructure of negative Poisson's ratio structures with respect to different design requirements. Multi-objective optimization problems involve optimizing a vector of objective functions and are often referred to as vector optimization problems. Typically, because the optimal solutions for different objectives in multi-objective optimization problems are generally not the same and can even be contradictory, achieving global optimality for all objectives simultaneously is usually unattainable. As a result, the solutions to multi-objective optimization problems often form a set, known as the Pareto set. Without loss of generality [24], the mathematical model for a multi-objective optimization problem can be described as follows:

$$\begin{aligned} &Minimize\, f_m(x),\ m = 1, 2, ..., M \\ &Subject\ to\ g_i(x) <= 0,\ i = 1, 2, ..., I \\ &\qquad H_j(x) = 0,\ j = 1, 2, ..., J \\ &X_k^{(L)} <= X_k <= X_k^{(U)},\ k = 1, 2..., n \end{aligned} \tag{3.10}$$

where M is the total number of subobjective functions, $f_m(x)$ is the mth subobjective function, I is the total number of inequality constraints, $g_i(x)$ is the ith inequality constraint, J is the total number of equation constraints, $H_j(x)$ is the jth equation constraint, n is the total number of design variables, X_k is the kth design variable, ${X_k}^{(L)}$ and ${X_k}^{(U)}$ are the lower and upper limits of the values taken by the kth design variable, respectively.

3.3.1 NSGA-II Multi-objective Optimization Algorithm

The foundation of Multi-objective Genetic Algorithms lies in Genetic Algorithms (GAs). Since the principles of genetic algorithms are based on population-based operations, and the goals of both multi-solution set and single-solution set optimization are consistent, using genetic algorithms to solve multi-objective problems is suitable. Moreover, there is no need for normalization; multi-objective optimization problems can be directly handled [25]. Multi-objective genetic algorithms do not independently seek Pareto optimal solutions. Instead, they find the Pareto front in a single run, which is effective in addressing complex Pareto fronts, including concave regions, making it suitable for solving multi-objective optimization problems.

The main features of the Non-Dominated Sorting Genetic Algorithm (NSGA) for multi-objective optimization include the use of non-dominated sorting and fitness sharing strategies. These strategies prevent the population from converging solely to a few individual solutions and help maintain diversity within the population [26]. NSGA-II, an improved version of the NSGA algorithm, enhances the capability of advancing the Pareto front by introducing the concept of an archive. It replaces fitness sharing with the concepts of crowding distance and crowding distance sorting. This helps in avoiding local concentration on the Pareto front and enables better exploration of the front. In the optimization of the double-arrowhead negative Poisson's ratio structure discussed in this paper, NSGA-II is employed as the multi-objective genetic algorithm. The primary steps of NSGA-II optimization design are as follows [27].

In the first step, the initial parent population P *is* randomly generated with size n. The objective function of multiple objectives of individuals is calculated, and then the population is ranked according to non-dominated individuals. The rank with order 1 is the best rank, the next best rank with order 2, and so on. The principle of ranking is that non-dominated individuals in the current population are assigned order 1 and removed from competition, and then non-dominated individuals are selected from the remaining current population and assigned order 2 to them until all individuals are assigned to the corresponding order. The value of fitness of each individual is equal to the order value, and the chromosome of order 1 is deposited into O. Also, the empty set Q is created and.

In the second step, two chromosomes C_i and C_j are selected in R, where the smaller the fitness of the individual, the higher the probability of being selected.

In the third step, a crossover operation is performed on C_i and C_j using a probability t_1, and they swap some genes with each other to generate two new chromosomes C_i' and C_j', and then a mutation operation is performed on C_i' and C_j' using a probability t_2, and the general mutation probability t_2 is much smaller than the swap probability t_1, , and the crossover and variation operations are performed using the SBX (Simulated Binary Crossover) crossover operator.

In the fourth step, the newly generated C_i' and C_j' undergo the calculation of their respective objective functions. Subsequently, C_i' and C_j' are added to the set Q. If the set Q contains fewer than n new chromosomes, the process returns to the second step for further iterations. This step helps ensure that the population size is maintained.

In the fifth step, the set $R = P \cup Q$ operation is performed, the geometry Q is emptied, the non-dominated individuals in the current set R are sorted, and then the set O is updated, the value of the iteration counter t *is* increased, and the optimization process is stopped if the iteration counter t *reaches a* predetermined value and the set O is returned, at which point the set O is the Pareto optimal solution set, and if it does not reach a predetermined value, it is transferred to the second step.

For a detailed description of the steps of the NSGA-II multi-objective genetic algorithm, you can refer to the source provided in reference.

3.3.2 *Multi-objective Optimization of Two-Dimensional Negative Poisson's Ratio Structure Design*

In the design of double-arrow type negative Poisson's ratio structure, there are different requirements for the structure from different aspects, and there are conflicts between different properties of the negative Poisson's ratio structure, usually each objective is constrained to each other, and to achieve one performance goal must sacrifice another performance goal as a prerequisite, so it is necessary to find a balance between different objectives of the structure and find an optimal solution set to meet the design requirements.

3.3.2.1 Optimization Objectives and Constraints

The double-arrowhead negative Poisson's ratio structure is known for its lightweight properties, energy absorption capabilities, and variable stiffness characteristics. An optimized design for this structure should maintain good performance while keeping its weight relatively low. In this study, the negative Poisson's ratio structure has minimal internal filler material, which typically results in lower stiffness. When applied to automotive components, this can lead to deformation under external forces. Therefore, in the optimization process, it's desirable for the negative Poisson's ratio structure to exhibit minimal deformation (d) in the vertical direction under external loads. The weight of the car body plays a critical role in overall fuel efficiency, emissions, and production costs. In the automotive design and manufacturing process, a lighter vehicle offers advantages, provided it maintains equivalent performance. Consequently, when the double-arrowhead negative Poisson's ratio structure is applied to automotive components, it's beneficial for the structure's weight (m) to be as low as possible. In the design of negative Poisson's ratio structures, it is essential to keep the stress level (S) at a low level to ensure durability and reduce plastic deformation. Taking all these factors into consideration, in this section's multi-objective optimization design process for two-dimensional double-arrowhead negative Poisson's ratio structures, deformation (d) and weight (m) are considered as the optimization objectives, while stress (S) is treated as a constraint, with a stress level less than 170MPa deemed to meet design requirements.

As the multi-objective genetic algorithm optimizes and generates multiple sets of Pareto non-dominated solutions, to find a solution set that meets the design requirements, this paper defines a satisfaction function C as in Eq. (3.11). From the definition of the satisfaction function, it is evident that smaller satisfaction values correspond to more optimal solutions.

$$C = \frac{d - d_{\min}}{d_{\max} - d_{\min}} + \frac{m - m_{\min}}{m_{\max} - m_{\min}} \tag{3.11}$$

where, d_{max} and d_{min} are the maximum and minimum values of the negative Poisson's ratio structural reference point displacement of the objective function in the Pareto optimal solution set; m_{max} and m_{min} are the maximum and minimum values of the negative Poisson's ratio structural mass of the double-arrow type of the objective function in the Pareto optimal solution set.

3.3.2.2 Optimization Results

In this section, when optimizing the microstructural parameters of the double-arrowhead negative Poisson's ratio structure to enhance its macroscopic performance, the first step is to establish a parameterized model for the structure. This model allows for the rapid generation of finite element analysis models with boundary load conditions by altering the geometric parameters of the structure. Following this, a Design of Experiments (DOE) methodology is used in conjunction with response surface methods to create response surface models for the parameterized structure. Finally, the NSGA-II multi-objective genetic algorithm is applied to perform multi-objective optimization on the model. Figure 3.5 depicts the optimization design process for the double-arrowhead negative Poisson's ratio structure in this chapter.

The efficiency and accuracy of obtaining an optimized solution set in the NSGA-II algorithm can be influenced by the choice of different parameters. Taking various influencing factors into account, this paper selects different optimization parameters for the NSGA-II algorithm, as shown in Table 3.6.

The initial design parameters of the negative Poisson's ratio microstructure are shown in Table 3.6. The selection of variables and the variation range of variables in the optimization process are consistent with the stiffness maximization optimization in Fig. 3.6. The NSGA-II multi-objective optimization algorithm has obtained 33 sets of Pareto non-inferiority optimal solution sets after optimization. As shown in Fig. 3.6, it can be seen that the multi-objective genetic algorithm NSGA-II can better provide a variety of alternative design solutions.

Table 3.7 lists five sets of Pareto optimization solutions for the two-dimensional double-arrow type negative Poisson's ratio structure. Comparing the different Pareto optimization solutions, it can be seen that there is a contradiction between the displacement in the vertical direction and the mass of the negative Poisson's ratio structure in the optimization objective of the double-arrow negative Poisson's ratio structure, and reducing the displacement of the negative Poisson's ratio structure in the optimization process usually comes at the cost of increasing the total mass of the structure, while reducing the mass of the structure usually increases the displacement of the negative Poisson's ratio structure in the vertical direction. Therefore, in the optimization process, if the design goal favors a higher stiffness of the structure, i.e., the minimum displacement of the negative Poisson's ratio structure, option 5 should be chosen; if the design goal favors the lightest mass of the negative Poisson's ratio structure, option 2 should be chosen.

Since the displacement and the mass of the structure in the vertical direction of the two-dimensional double-arrow type negative Poisson's ratio structure cannot be

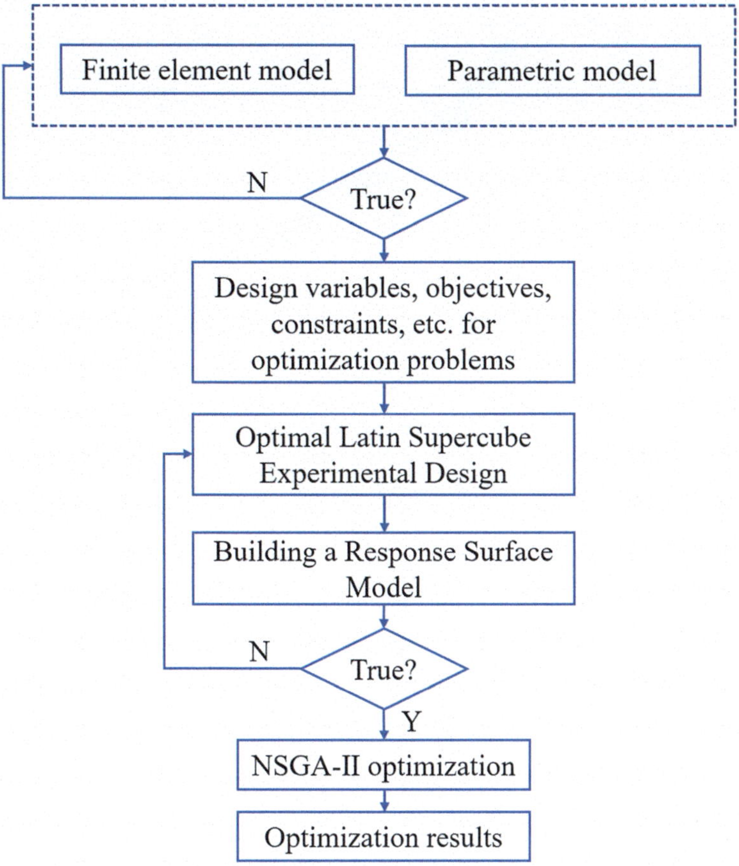

Fig. 3.5 Optimal design flow of negative Poisson's ratio structure

Table 3.6 NSGA-II optimized parameter design

Parameters	Values
Population size	200
Evolutionary algebra	400
Cross probability	0.87
Cross distribution index	15
Variation distribution index	80

optimized at the same time, this paper solves the satisfaction of each Pareto optimization solution according to the satisfaction function defined in Eq. (3.11), and selects solution 4 in Table 3.8 as the satisfactory solution, and the optimization results are compared with the optimized initial values in Table 3.8. From Table 3.8, it can be seen that the two-dimensional double-arrow type negative Poisson's ratio structure After

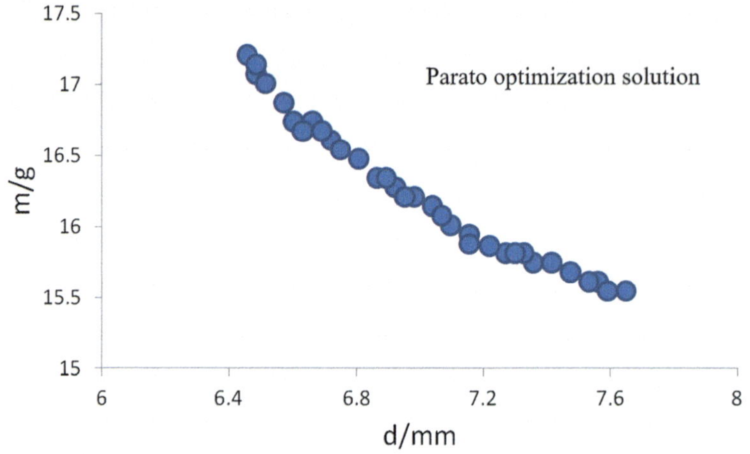

Fig. 3.6 Pareto non-inferior optimal solution set

Table 3.7 Comparison of Pareto optimized solutions

Design variable	Option 1	Option 2	Option 3	Option 4	Option 5
l_l/mm	19.27	16.36	17.64	20.37	22.37
θ/rad	0.36	0.41	0.28	0.36	0.28
φ/rad	0.39	0.62	0.54	0.64	0.51
a/mm	7.10	8.82	8.92	7.56	6.87
T_1/mm	2.34	2.34	2.64	1.93	1.79
T_2/mm	1.14	0.86	0.72	1.02	0.94
d/mm	7.01	7.61	7.48	6.79	6.43
m/g	16.18	15.24	15.89	16.42	17.09

the multi-objective optimization, the mass of the structure is significantly reduced and the displacement under the external force becomes smaller. The comparison of the negative Poisson's ratio structure before and after optimization is shown in Fig. 3.7.

In order to compare and verify the effect of multi-objective optimization of the negative Poisson's ratio structure of double-arrow type, the multi-objective optimization results are compared with the single-objective optimization results respectively. The single-objective structural optimization is based on the genetic algorithm for single-objective optimization of the vertical displacement of the negative Poisson's ratio structure with weight and stress as constraints and for single-objective optimization of the weight of the structure with displacement and stress as constraints, respectively. The comparison of different optimization results is presented in Table 3.9.

Table 3.8 Comparison table of before and after optimization of multi-objective genetic algorithm

Design variable	Initial	NSGA-II
l_l/mm	18.50	20.37
θ/rad	0.34	0.36
φ/rad	0.56	0.64
a/mm	8.50	7.56
T_1/mm	2.50	1.93
T_2/mm	0.80	1.02
d/mm	7.89	6.79
m/g	17.63	16.42

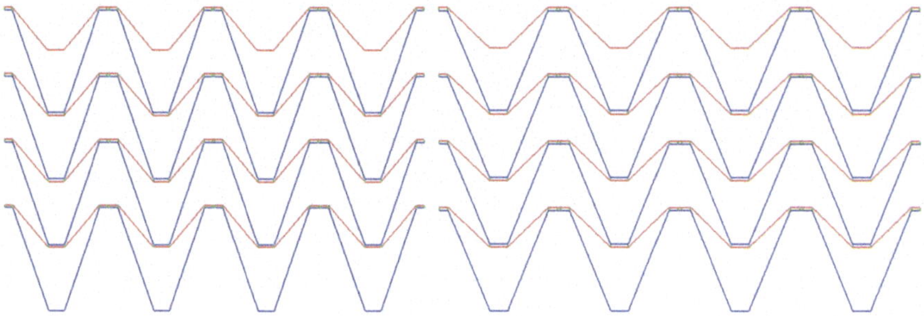

(a) Original negative Poisson's ratio structure

(b) NSGA-II optimized negative Poisson's ratio structure

Fig. 3.7 Comparison of negative Poisson's ratio structure before and after optimization

Table 3.9 Comparison table of optimization results of multi-objective genetic algorithm and single-objective optimization algorithm

Design variable	Single objective optimization value_m	Single objective optimization value_d	NSGA-II
l_l/mm	16.86	22.45	20.37
θ/rad	0.29	0.41	0.36
φ/rad	0.54	0.68	0.64
a/mm	8.9	7.27	7.56
T_1/mm	2.37	2.24	1.93
T_2/mm	0.95	0.64	1.02
d/mm	8.46	5.94	6.79
m/g	14.79	19.94	16.42

Comparing the optimization results, when performing single-objective optimization with the goal of reducing the weight of the negative Poisson's ratio structure, the weight decreased to 14.79 g, a 16.11% reduction from the initial design. However, under external forces, the vertical displacement increased to 8.46 mm, a 7.22% increase compared to the initial design. On the other hand, when performing single-objective optimization with the goal of reducing the vertical displacement of the negative Poisson's ratio structure, the displacement was reduced to 5.94 mm, a 24.71% decrease from the initial design. However, the weight of the structure increased to 19.94 g, a 13.10% increase compared to the initial design. In contrast, multi-objective optimization using the NSGA-II algorithm achieved a displacement of 6.79 mm, a 13.94% reduction from the initial design, and a structure weight of 16.42 g, a 6.86% reduction from the initial design. This approach effectively balanced the reduction of vertical displacement and the decrease in structure weight, achieving a good compromise between the two different optimization objectives. In summary, when single-objective optimization targets weight reduction, it effectively achieves lightweight design but results in higher vertical displacement. When single-objective optimization targets reducing vertical displacement, it reduces the displacement but significantly increases the weight. Multi-objective optimization based on the NSGA-II algorithm effectively reduces the vertical displacement while also lowering the structure's weight, striking a good balance between the two distinct optimization objectives.

3.3.3 Multi-objective Optimization Design of Three-Dimensional Negative Poisson's Ratio Structure

Based on the analysis in the first chapter of the document, it is evident that three-dimensional negative Poisson's ratio structures have controllable microstructure regularity and more pronounced negative Poisson's ratio characteristics. Therefore, compared to two-dimensional structures, three-dimensional negative Poisson's ratio structures are more suitable for practical engineering applications. These structures, particularly the double arrowhead design, exhibit excellent energy absorption properties and can be used in energy-absorbing components to improve their performance. In this section, the optimization process outlined in Fig. 3.7 is applied to a three-dimensional double arrowhead negative Poisson's ratio structure, which consists of $4 \times 7 \times 24$ individual cells, as shown in Fig. 3.8.

To investigate the energy absorption characteristics of the negative Poisson's ratio structure, this subsection equates it according to the national automotive low-speed crash regulations by fixing six degrees of freedom at one end of the structure in Ls-Dyna, and then impacting the other end of the structure with a rigid wall of 800 kg at 16 km/h. Specific energy absorption (SEA) is an important indicator of the energy

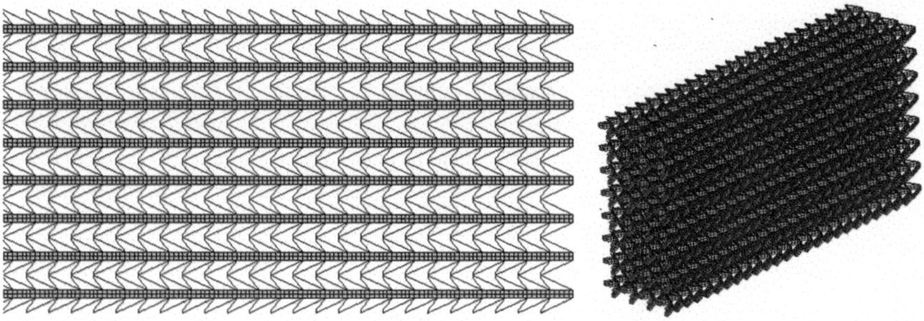

Fig. 3.8 Three-dimensional negative Poisson's ratio structural model

absorption performance of a structure, and the specific energy absorption of a structure should generally be maximized in the design process. The maximum collision force (PCF) is another important index for the evaluation of the crashworthiness of the structure. The maximum collision force is closely related to the chance of casualties of the members during the collision, and an excessive maximum collision force will cause serious casualties, so the maximum collision force value should be strictly controlled during the collision. The weight of the negative Poisson's ratio structure has an important impact on its cost control and performance, so it should also be considered in the design process. In this subsection, the maximum collision force and specific energy absorption are jointly used as the optimization objectives, and the weight of the structure is used as the constraint when the 3D negative Poisson's ratio structure is optimally designed. According to the variation range of the parameters of the negative Poisson's ratio structure, the mathematical model of the optimized design can be expressed as follows.

$$\begin{gathered} \text{Minimize } 1/\text{SEA},\ \text{PCF} \\ \text{Subjected to :} \\ \text{m} \leq 0.35 \\ 0 < \theta < \varphi < 1.57 \\ 1 < \text{a} < 15, 5 < \text{l}_1 < 30 \\ 0.1 < \text{T}_1 < 4, 0.1 < \text{T}_2 < 4 \end{gathered} \tag{3.12}$$

By combining parametric modeling, the Design of Experiments (DOE) method, Response Surface Methodology (RSM), and the NSGA-II multi-objective genetic algorithm, the optimization process's parameter design is consistent with the multi-objective optimization of the two-dimensional negative Poisson's ratio structure. After 32 iterations, the NSGA-II multi-objective optimization algorithm obtained a set of 27 Pareto non-dominated optimal solutions. Table 3.10 presents five groups of Pareto optimal solutions for the three-dimensional double arrow negative Poisson's ratio structure.

Comparing the different optimized solution sets in the above table, it can be seen that the three-dimensional double-arrow type negative Poisson structure with

Table 3.10 Comparison of Pareto-optimized solutions for three-dimensional negative Poisson's ratio structures

Design variable	Original scheme	Option 1	Option 2	Option 3	Option 4	Option 5
l_l/mm	14	13.5	15.2	11.6	12.0	17.1
θ/rad	0.3	0.47	0.37	0.43	0.42	0.28
φ/rad	0.5	0.61	0.56	0.58	0.63	0.49
a/mm	2	2.1	1.6	1.7	2.5	3.0
T_1/mm	1	1.2	1.5	1.4	0.9	0.7
T_2/mm	0.8	0.4	0.7	1.3	0.5	0.6
SEA/J/g	11,893	12,861	12,055	14,219	13,764	14,892
PCF/kN	82.7	67.2	62.6	75.8	71.9	80.4

different performance can be obtained by different parameter designs, and there are inconsistencies between the optimized target specific energy absorption and maximum collision force in the collision process, and the optimized structure should be selected according to the specific design requirements during the design. In the design process, if more attention is paid to the specific energy absorption of the structure, option 5 should be selected, and if more attention is paid to the maximum collision force of the structure, option 2 is more suitable. Since the maximum collision force and specific absorption energy of the three-dimensional double-arrow type negative Poisson's ratio structure cannot be optimized simultaneously in the optimization process, this subsection finds the satisfaction of each Pareto optimization solution according to the satisfaction function (3.11), and it can be found that among 27 different sets of Pareto optimization solutions, option 5 has the smallest satisfactory solution, so option 5 is the optimal solution set selected according to the satisfaction function. Comparing the original scheme with scheme 5, it can be seen that after the multi-objective optimization of the microstructure design parameters, the specific energy absorption of the three-dimensional double-arrow negative Poisson's ratio structure is significantly improved, and the maximum collision force during the collision is effectively reduced. For the application of the double-arrow negative Poisson's ratio structure, the three-dimensional double-arrow negative Poisson's ratio structure obtained in this section based on the NSGA-II multi-objective optimization will be applied as the initial design in the negative Poisson's ratio structure energy absorption box in Chap. 5, so as to effectively improve the performance of the structure.

From the analysis above, it's evident that by effectively combining the double arrow negative Poisson's ratio structure's parameterized model, response surface model, and NSGA-II multi-objective genetic algorithm, it is possible to perform multi-objective optimization on various aspects of the structure, including weight, vertical displacement, Specific Energy Absorption (SEA), and Peak Contact Force (PCF). This optimization ensures that the negative Poisson's ratio structure continues to meet design requirements while improving its performance through the optimization of topology, shape, and thickness. This approach leads to the enhancement of the structure's properties.

3.4 Robustness Optimization of Negative Poisson's Ratio Structure

(1) Establish a parameterized model for the electric vehicle battery protection system. Select the shape parameters of the negative Poisson's ratio structure's core unit, the thickness, and cross-sectional parameters of the protective cross-beams and protective longitudinal beams as design variables to create a parameterized model for the electric vehicle battery protection system. By altering the design parameters, you can quickly create finite element analysis models for the battery protection system, enabling rapid performance analysis and optimization. The shape parameters for the core units of the negative Poisson's ratio structure mainly include the thickness, length, width, and the angle between the cell walls. The cross-sectional parameters for the protective crossbeams and longitudinal beams primarily consist of length, width, and height parameters for different cross-sectional shapes.

(2) Optimize the positions and shapes of the front and rear induction slots. Analyze the energy absorption performance of the battery protection system in front and rear collision scenarios based on automotive crash conditions. To enhance the system's ability to absorb energy during collisions and reduce the impact on the battery, add induction slots to the front and rear protective crossbeams. Design the positions, widths, and depths of these induction slots within the parameterized model, and then use genetic algorithms for optimization. This process allows you to design front and rear induction slots that maximize the energy absorption capability of the battery protection system.

(3) Establish an approximate model for the electric vehicle battery protection system. Based on the parameterized model of the battery protection system, along with the Latin hypercube sampling method and quadratic response surface methodology, create a computational approximate model for the electric vehicle battery protection system. To ensure a more even distribution of sample points and enhance the accuracy of the approximate model, employ the Latin hypercube sampling method for the parameter sample points. The design parameters primarily encompass the thickness, length, width, and the angle between the cell walls of the negative Poisson's ratio structure core unit, as well as the dimensions, width, height, and thickness of different cross-sectional shapes in the protective crossbeams and protective longitudinal beams. After obtaining sample data through Latin hypercube sampling, utilize the second-order response surface model method to establish an approximate model for the

relationship between the design parameters and the performance of the electric vehicle battery protection system.

(4) Verify the accuracy of the approximate model. After constructing the approximate model, validate its precision using three different methods: the coefficient of determination (R-squared), relative maximum absolute error, and mean absolute deviation. The approximate model's accuracy is deemed satisfactory when it meets the precision requirements in all three methods.

$$R^2 = 1 - \frac{\sum_{i=1}^{q} (y_i - \hat{y}_i)^2}{\sum_{i=1}^{q} (y_i - \overline{y})^2} \tag{3.13}$$

$$RMAE = \frac{\max\{|y_i - \hat{y}_i|, \ldots, |y_i - \hat{y}_q|\}}{\sum_{i=1}^{q} |y_i - \overline{y}| / q} \tag{3.14}$$

$$RAAE = \frac{\sum_{i=1}^{q} |y_i - \hat{y}_i|}{\sum_{i=1}^{q} |y_i - \overline{y}|} \tag{3.15}$$

where, $\hat{y}_i$ is the predicted value of the system response surface model, y_i is the real response value of the system response, which is obtained by finite element analysis, q is the number of sample points, and $\overline{y}$ is the average of the real response value. y_i

If the response surface model does not meet the required accuracy, return to step (3) to extract more sample points and improve the response surface model. If the response surface model meets the accuracy requirements, proceed to step (5).

(5) Establish a multi-objective robust optimization model for the electric vehicle battery protection system, considering Six Sigma standards and the system's optimization objectives, design variables, and constraints. The formulated multi-objective robust optimization model can be represented as follows:

$$\begin{cases} \min \{f_1(y_{\mu 1}(x), y_{\sigma 1}(x)), f_2(y_{\mu 2}(x), y_{\sigma 2}(x)), \ldots, f_n(y_{\mu m}(x), y_{\sigma m}(x))\} \\ s.t.\ C_{\mu j}(x) + 6C_{\sigma j}(x) \le 0 \\ x_L + 6x_\sigma \le x_\mu \le x_U - 6x_\sigma \end{cases} \tag{3.16}$$

where m is the number of optimization objectives, f_m is the $m\text{-}th$ optimization objective of the EV battery protection system, $y_{\mu m}(x)$, x_μ and $C_{\mu j}$ are the optimization objectives, design variables and constraints respectively, $y_{\sigma m}(x)$, x_σ and $C_{\sigma j}$ are the standard deviations of the optimization objectives, design variables and constraints respectively, x_U and x_L are the upper and lower bounds of the design variables respectively.

(6) Optimize the design parameters of protective crossbeams, protective longitudinal beams, and the negative Poisson's ratio structure's core unit for the electric vehicle battery protection system. When optimizing the design parameters of

the battery protection system, it's essential to strike a balance between different objectives, such as various collision scenarios (front and rear collisions), energy absorption by the battery protection system, and the system's weight. Leveraging the approximate model of the electric vehicle battery protection system established in the previous step and considering the system's performance under different collision conditions, use the NSGA-II multi-objective genetic algorithm to perform multi-objective optimization of the design parameters of the electric vehicle battery protection system under various scenarios.

(7) Select the optimal design for the electric vehicle battery protection system and validate it. To strike a balance between different optimization objectives, the NSGA-II multi-objective genetic algorithm will provide multiple sets of Pareto optimal solutions. During the design process, the selection should be based on design requirements. In this method, a satisfaction-based optimal solution selection approach is proposed, defined as follows:

$$C = \frac{(1/E) - (1/E)_{\min}}{(1/E)_{\max} - (1/E)_{\min}} + \frac{m - m_{\min}}{m_{\max} - m_{\min}} \tag{3.17}$$

$$E = \frac{1}{2}E_f + \frac{1}{2}E_b \tag{3.18}$$

where, E is the calculated value of energy absorbed by the battery protection system in front collision and rear collision, E_f is the energy absorbed by the battery protection system in front collision, E_b is the energy absorbed by the battery protection system in rear collision, $m_{\max}$, $m_{\min}$ are the maximum and minimum values of the mass of the battery protection system in the set of Pareto optimization solutions. When selecting the optimal design using the satisfaction-based optimal solution selection approach, the smaller the value of C, the better the solution set is considered to be, as it better meets the design requirements. After selecting the optimal solution, the parametric model is used to quickly create a finite element analysis model for front and rear collision analyses, thereby validating the optimized design (Fig. 3.9).

3.5 Gradient Optimization of Negative Poisson's Ratio Structure

The gradient-type negative Poisson's ratio structure mainly includes thickness gradient-type and angle gradient-type negative Poisson's ratio structures. The optimization analysis problem for thickness gradient-type negative Poisson's ratio structures can be summarized as an optimization problem with the structure's dimensions as design variables, structural shape, strength, and stiffness as constraints, and minimizing weight as the optimization objective. There are many intelligent optimization methods available, each with its own advantages and disadvantages. In this paper, genetic algorithms and particle swarm optimization algorithms are proposed

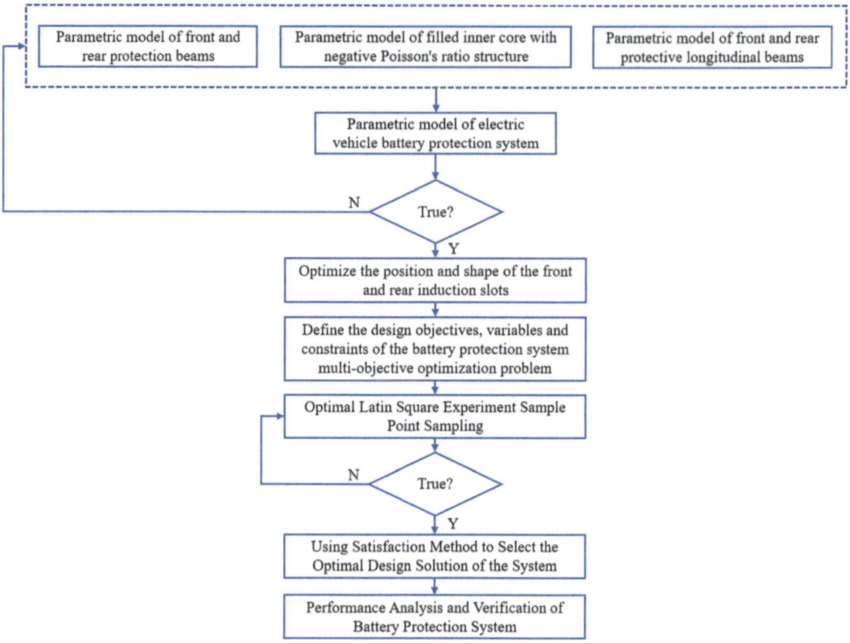

Fig. 3.9 Optimization flow chart

to perform optimization analysis on thickness gradient-type negative Poisson's ratio structures.

3.5.1 Thickness Gradient Type Negative Poisson's Ratio Structure and Design Parameters Selection

The geometric model of the planar thickness gradient-type negative Poisson's ratio structure used in this paper is shown in Fig. 3.10. The planar thickness gradient-type negative Poisson's ratio structure consists of 4 unit cells in the x-direction and 6 unit cells in the y-direction. The lengths in the x-direction and y-direction vary according to the optimization variables, and the height in the z-direction is 20 mm.

The thickness gradient negative Poisson's ratio structure is a structural form where the wall thickness of the negative Poisson's ratio structure's unit cells changes along a certain direction. This thickness gradient structure gradually alters the elastic properties of the negative Poisson's ratio structure in the gradient direction. Therefore, gradient negative Poisson's ratio structures, similar to equally strong step beam structures, offer advantages such as increased material utilization and reduced structural weight.

(a) Isometric View (b) Front view

Fig. 3.10 Planar negative Poisson's ratio structure

The finite element model of the thickness gradient type concave hexagonal type negative Poisson structure is modeled by a four-node shell cell SHELL163, and the whole structure is divided into 35,280 cells and 36,247 nodes. Each layer of monolithic cells is given a different thickness, and the thickness of the joint between two layers of monolithic cells is the sum of the thicknesses of the two layers of monolithic cells, as shown in Fig. 3.11.

In this paper, we take aluminum alloy as an example to study the thickness gradient type concave hexagonal negative Poisson structure under torsional load.

The optimum structure for the conditions. The physical properties of the aluminum alloy are shown in Table 3.11.

In the optimization analysis process, the design variables are continuously changing. Therefore, before establishing the model, it is necessary to select geometric parameters that accurately represent the structure's form. The chosen geometric parameters should be independent and comprehensive in representing the structure's form. This paper uses the APDL (ANSYS Parametric Design Language) in the

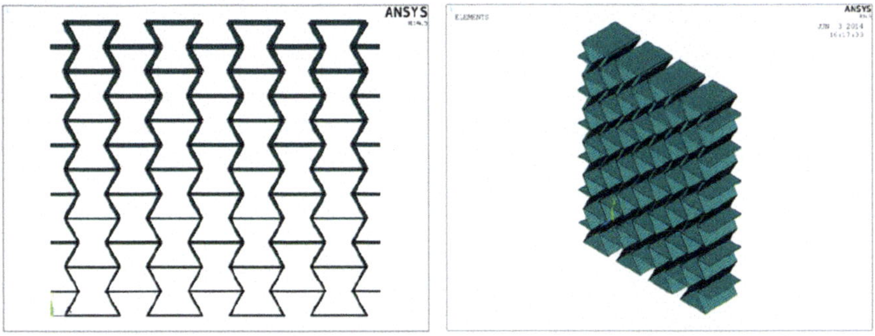

(a) Front view (b) Isometric View

Fig. 3.11 Finite element model of negative Poisson's ratio structure with thickness gradient

Table 3.11 Material properties

Mechanical property parameters	Value
Tensile modulus of elasticity (MPa)	70,000
Poisson's ratio	0.3
Density (kg/mm^3)	2.7×10^{-6}

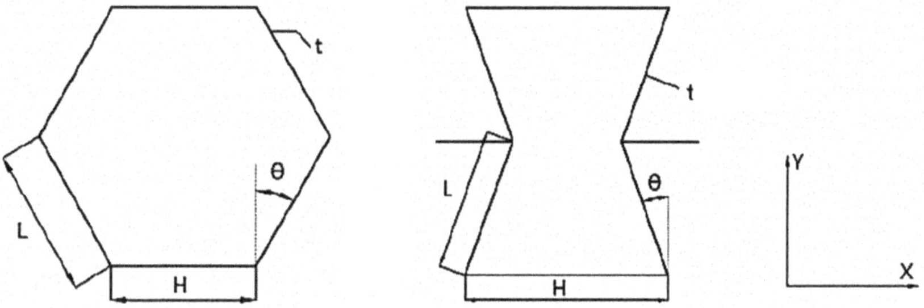

Fig. 3.12 Geometric parameters of the single cell structure of positive and negative honeycomb structure

ANSYS finite element analysis software to create a parametric model of the thickness gradient concave hexagonal negative Poisson's ratio structure. The advantage of using ANSYS's APDL is that it allows modeling with parameters instead of numerical inputs for model dimensions, material properties, and parameterized meshing. It also provides the capability to extract stress, displacement, and other information from the ANSYS database, ensuring that the structure's shape remains consistent during the optimization process and ensuring the convergence of the finite element analysis. As shown in Fig. 3.12, it depicts the geometric parameters of the positive and negative hexagonal cell structure.

Based on the geometric characteristics of the model, there are five optimization parameters for the thickness gradient concave hexagonal negative Poisson's ratio structure: initial thickness t_1, gradient index tt, transverse cell wall length L, ratio of longitudinal cell wall length to transverse cell wall length H/L, and concave angle θ, totaling five parameters. Since the shape of the optimized model remains unchanged during the optimization process, the optimal design of the thickness-gradient concave hexagonal negative Poisson structure is dimensional optimization.

3.5.2 *Negative Poisson's Ratio Structural Constraints and Loading*

This paper primarily investigates the mechanical response of the structure under torsional loads. One end of the structure is fixed, and torsional loads are applied at the

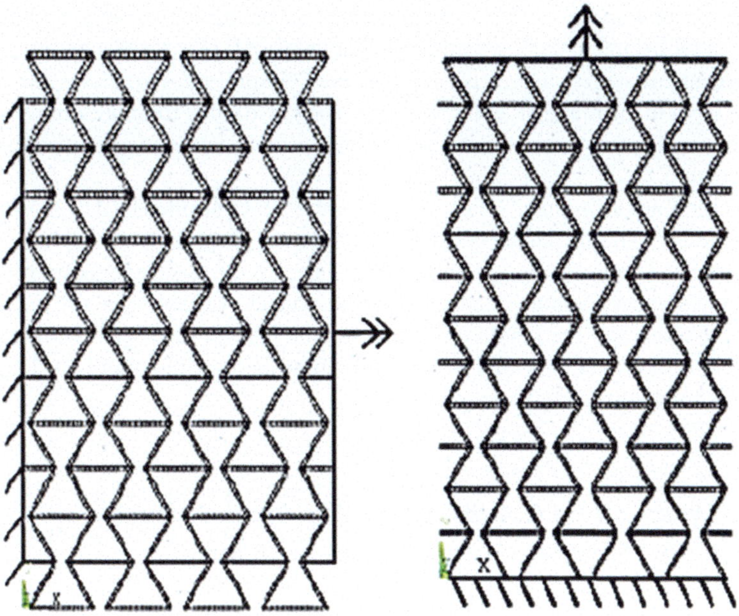

(a) X direction torsion (b) Y direction torsion

Fig. 3.13 Load and boundary conditions in different directions

other end. To comprehensively study the torsional performance of the gradient negative Poisson's ratio structure and achieve better optimization results, this paper applies loading in different directions. The structural constraints and loading conditions are depicted in Fig. 3.13.

Constraints refer to the range of values for various design variables and state variables, with the objective of minimizing the weight of the structure. In order to compare it with the positive honeycomb structure, the range of the concave angle has been expanded. The allowable ranges for the five design variables in this chapter are presented in Table 3.12.

Table 3.12 Value range of design variables

Design variable	Minimum	Maximum
T1	0.5	2
T2	0	0.05
L	10	30
H/L	2	4
θ	−60 (Positive honeycomb)	60 (negative poisson's ratio)

To compare with the uniform case, the range of the design variable, the gradient index tt is set from 0 to 0.05, where a gradient index of 0 represents a uniform structure. To compare with both positive and negative Poisson's ratio structures, the design variable is set in the range of −60 to 60, with a negative value for representing a positive honeycomb structure and a positive value for representing a negative Poisson's ratio structure.

State variables are typically the dependent variables controlled by design variables and constitute a significant aspect of the constraints. They generally include stress, displacement, temperature, and so on. This paper focuses on the static response of structures under torsional loads, so the chosen state variables are the maximum equivalent stress and the maximum displacement. This ensures that the maximum equivalent stress remains within the strength limits and the maximum displacement falls within the allowable displacement.

3.5.3 Optimization Results for the Thickness Trapezoid of the Negative Poisson's Ratio Structure

Genetic Algorithm Results for the x-Direction Torsional Thickness Gradient Negative Poisson's Ratio Structure.

Under the given genetic algorithm parameter settings, after 16 iterations, the optimal values of the design variables are obtained, as shown in Eq. (3.19). The optimized design variables are as follows, according to Eq. (3.20):

$$\begin{cases} T = 50 \\ N = 100 \\ pm = 0.4 \\ pc = 0.01 \end{cases} \tag{3.19}$$

where

T the maximum number of iterations of the stopping condition.
N number of populations.
pm crossover rate.
pc variation rate.

$$X = \left[X_1\ X_2\ X_3\ X_4\ X_5 \right] = \left[t_1\ tt\ l\ h/l\ \theta \right] = \begin{bmatrix} 0.696 \\ 0.05 \\ 10 \\ 2 \\ 60 \end{bmatrix}^T \tag{3.20}$$

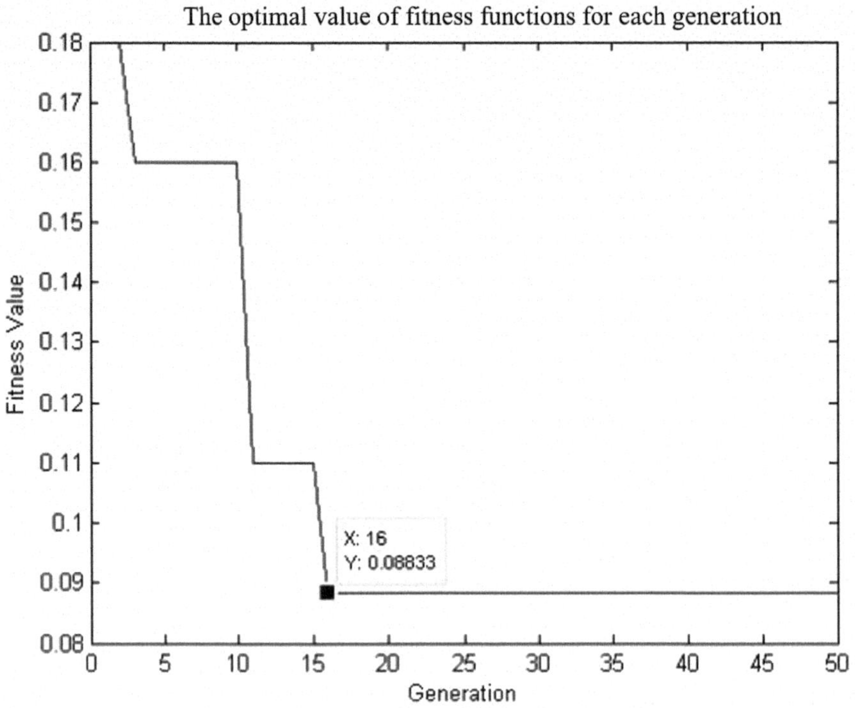

Fig. 3.14 Optimal values of the fitness function for each generation

The best fitness value, which corresponds to the minimum weight of the structure, is f(X) = 0.08833 kg. The convergence of the fitness function is shown in Fig. 3.14.

The results of negative Poisson's ratio structure of torsion thickness gradient in y direction by genetic algorithm.

After iterating 35 generations, the optimal optimization variable value is obtained, as shown in Eq. (3.21).

$$X = \begin{bmatrix} X_1 \; X_2 \; X_3 \; X_4 \; X_5 \end{bmatrix} = \begin{bmatrix} t_1 \; tt \; l \; h/l \; \theta \end{bmatrix} = \begin{bmatrix} 0.5 \\ 0 \\ 10 \\ 2 \\ -60 \end{bmatrix}^T \tag{3.21}$$

The best fitness value, which corresponds to the minimum weight of the structure, is f(X) = 0.054 kg. The convergence of the fitness function is illustrated in Fig. 3.15.

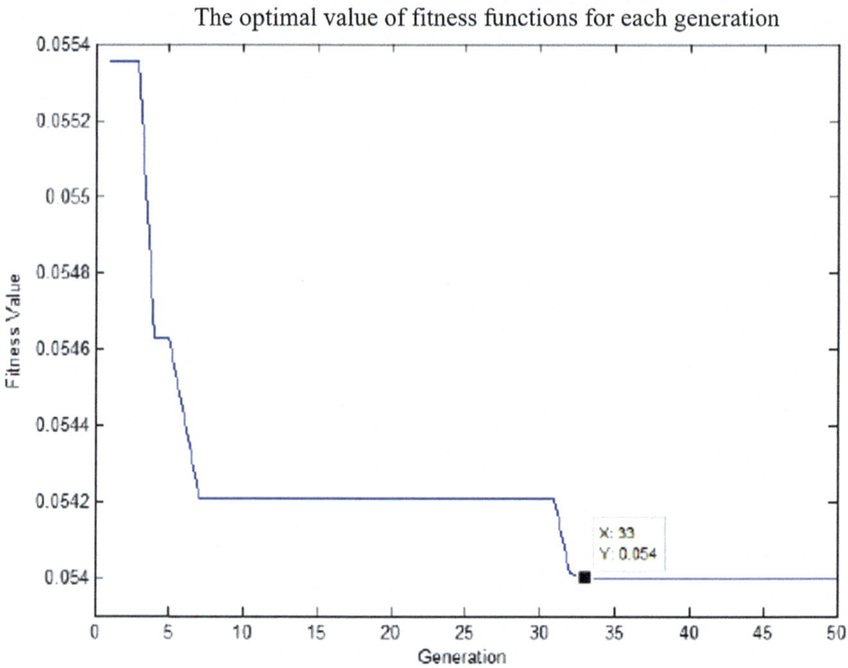

Fig. 3.15 Optimal values of the fitness function for each generation

3.6 Summary of This Chapter

In this chapter, a systematic study was conducted on the optimization of the microstructural parameters of negative Poisson's ratio structures to enhance their macroscopic performance. Building upon the application of parameterized modeling in general structural optimization, a parameterized model for the double arrowhead negative Poisson's ratio structure was established using MATLAB programs. This parameterized model enables the rapid creation of different shapes of negative Poisson's ratio structure models, greatly improving the efficiency and accuracy of analysis and optimization for such structures. A methodology was proposed, which combines the parameterized model of negative Poisson's ratio structures, optimal Latin hypercube experimental design, response surface methodology, and optimization algorithms to achieve the optimization of double arrowhead negative Poisson's ratio structures. This approach facilitates the rapid optimization of negative Poisson's ratio structure performance and enhances the efficiency of their optimization designs. Research findings demonstrate that the optimization methods introduced in this chapter successfully address various design requirements for double arrowhead negative Poisson's ratio structures, allowing for single-objective and multi-objective parameter optimization. This approach achieves the optimal topology, shape, and thickness design of double arrowhead negative Poisson's ratio structures, laying a solid foundation for the application of these structures in the automotive industry.

References

1. Ankita Jaisingh Sheoran, Harish Kumar. Fused Deposition modeling process parameters optimization and effect on mechanical properties and part quality: Review and reflection on present research [J]. Materials Today: Proceedings, 2020, 21(Pt3):1659–1672.
2. Chetan A. Somani, D.I. Lalwani. Experimental Investigation of TIG-MIG Hybrid Welding Process on Austenitic Stainless Steel [J]. Materials Today: Proceedings, 2019, 8(Pt7):4826–4834.
3. Montgomery, D. C. (2008). Design and analysis of experiments. John Wiley & Sons.
4. Ellison, S. L., Grimes, N., Fardus, F., and Ellison, M. P. (2012). Use of a replicated latin square design in a homogeneity test for high purity organic melting point standards. Accreditation and Quality Assurance, 17(3):283–290.
5. Zheng Y, Das PK. Improved response surface method and its application to stiffened plate reliability analysis [J]. Engineering Structures, 2000(22):544–551.
6. Roussouly N, Petitjean F. A new adaptive response surface method for reliability analysis [J]. Probabilistic Engineering Mechanics, 2013(32):103–116.
7. Li J, Wang H. Doubly weighted moving least squares and its application to structural reliability analysis [J]. Structural and Multidisciplinary Optimization. 2012(46):69–82.
8. Goswami S, Ghosh S. Reliability analysis of structures by iterative improved response surface method [J]. Structural Safety, 2016(60):56–66.
9. Ziwu Ren and Ye San. A Hybrid Optimized Algorithm Based on Simplex Method and Genetic Algorithm [C]. 2006 6th World Congress on Intelligent Control and Automation, 2006: 3547–3551.
10. Y. Gao and D. Song. A New Improved Genetic Algorithms and its Property Analysis [C]. 2009 Third International Conference on Genetic and Evolutionary Computing, 2009: 73–76.
11. Yi-nan Guo, Dun-wei Gong and Zhen-gui Xue. Hybrid Optimization Method Based on Genetic Algorithm and Cultural Algorithm [C]. 2006 6th World Congress on Intelligent Control and Automation, 2006: 3471–3475.
12. Echterr, Bischoffm. Numericalefficiency, locking and unlocking of NURBS finite elements [J]. Computer Methods in Applied Mechanicsand Engineering, 2010, 199(5–8):374–382.
13. Hughestj, Cottrellja, Bazilevsy. Isogeometric analysis: CAD, finite elements, NURBS, exact geometry and mesh refinement [J]. Computer methods in applied mechanics and engineering, 2005, 194(39–41):4135–4195.
14. Ding C, Cui X, Deokar R R, et al. An isogeometric independent coefficients (IGA-IC) reduced order method for accurate and efficient transient nonlinear heat conduction analysis [J]. Numerical Heat Transfer, Part A: Applications, 2018, 73(10):667–684.
15. Kangh, Chenf, Liy, et al. Knot calculation for spline fitting via sparse optimization [J]. Computer-Aided Design, 2015, 58:179–188.
16. Kapl M, Buchegger F, Bercovier M, et al. Isogeometric analysis with geometrically continuous functions on planar multi-patch geometries [J]. Computer Methods in Applied Mechanics and Engineering, 2017, 316:209–234.
17. Gao Q, Wang L, Zhou Z, Ma Z D, Wang C, Wang Y. Theoretical, numerical and experimental analysis of three-dimensional double-V honeycomb [J]. Materials and Design, 2018, 139: 380–391.
18. Fan Z, Xia L, Lai W, et al. Evolutionary topology optimization of continuum structures with stress constraints [J]. Structural and Multidisciplinary Optimization, 2019, 59:647–658.
19. Ho-Nguyen-Tan T, Kim H G. Level set-based topology optimization for compliance and stress minimization of shell structures using trimmed quadrilateral shell meshes [J]. Computers & Structures, 2022, 259:106695.

20. Luo Y, Wang MY, Kang Z. An enhanced aggregation method for topology optimization with local stress constraints. Computer Methods in Applied Mechanics and Engineering, 2013, 254: 31–41.
21. Rong JH, Xiao TT, Yu LH, et al. Continuum structural topological optimizations with stress constraints based on an active constraint technique. International Journal for Numerical Methods in Engineering, 2016, 108(4): 326–360.
22. Hirsch, Jürgen. Aluminium in Innovative Light-Weight Car Design [J]. Materials Transactions, 2011, 52(5):818–824.
23. R. Yang, Y. Fu, G. Li. Application of Tailor Rolled Blank in Vehicle Front End for Frontal Impact [C]. SAE Technical Paper, 2007–01–0675, 2007.
24. Murata T, Ishibuchi H. MOGA: Multi-objective genetic algorithms [C]. IEEE international conference on evolutionary computation, Perth, WA, Australia, 1995, 1:289–294.
25. Deb K, Pratap A, Agarwal S, et al. A fast and elitist multiobjective genetic algorithm: NSGA-II [J]. IEEE transactions on evolutionary computation, 2002, 6(2):182–197.
26. Yuhan X, Qing G, Aopin T, et al. Multi-objective Route Planning of Museum Guide based on an Improved NSGA-II Algorithm [J]. Journal of Physics: Conference Series, 2021, 1828(1).
27. Luca C, Vittorio F. Improved NSGA-II algorithms for multi-objective biomarker discovery [J]. Bioinformatics (Oxford, England), 2022, 38(Supplement_2).

Chapter 4
Manufacturing Method of Structures with Negative Poisson's Ratio

4.1 Frontiers

More than 160 years ago, the possibility of negative Poisson's ratio materials was already proposed by the classical elasticity theory. Early research has found that certain natural materials exhibit negative Poisson's ratio characteristics, including molecular materials with special structures and biological materials with fibrous or porous microstructures, such as pyrite crystals, cubic metal elements, porous cancellous bone, and cat skin. With the continuous advancement of research on negative Poisson's ratio materials, most of them are currently artificially designed and manufactured.

Metallic negative Poisson's ratio materials and structures refer to materials and structures that have negative Poisson's ratio characteristics and have metal as the base material. They mainly include metals with negative Poisson's ratio microstructures, negative Poisson's ratio foam metals, and negative Poisson's ratio objects made with metal as the base material. Zhang et al. used theoretical analysis and experiments to investigate the tensile properties of iron-gallium alloys and iron-aluminum alloys, and used density functional theory to determine the negative Poisson's ratio effects of the two alloys. Mitschke et al. used the same method to manufacture a titanium alloy Ti–6Al–4V wire elastic porous structure, and measured its wire elasticity through tensile tests, observing a negative Poisson's ratio value of −0.75. Dirrenberger et al. conducted a full stress field simulation of negative Poisson's ratio metal microstructures and found that the negative Poisson's ratio effect of the structure continued to exist and became stronger in the plastic yield stage. Taylor et al. experimentally demonstrated that low porosity periodic perforated metal structures can exhibit negative Poisson's ratio characteristics using aluminum honeycomb panels. Ren et al. developed new methods for designing three-dimensional metallic negative Poisson's ratio metamaterials and negative Poisson's ratio tubular structures. Compared with traditional negative Poisson's ratio materials, three-dimensional metallic negative

G. Zhou et al., *Design, Manufacturing, and Application of Negative Poisson's Ratio Structure*, https://doi.org/10.1007/978-981-96-1937-5_4

Poisson's ratio metamaterials and three-dimensional negative Poisson's ratio tubular structures have excellent mechanical properties.

Negative Poisson's ratio materials composed of multiple ordinary base materials are called multiple (multimaterial) negative Poisson's ratio materials. Wang et al. designed and manufactured some dual negative Poisson's ratio metamaterials using computer-aided design and 3D printing technology. Their mechanical properties are significantly different from those of negative Poisson's ratio materials composed of a single material. With the introduction of another material, the mechanical properties of the metamaterials are greatly improved without changing the overall geometric structure of the negative Poisson's ratio cells. Vogiatzis et al. used topology optimization methods to design and manufacture negative Poisson's ratio materials with multiple physical properties.

Existing research on composite negative Poisson's ratio materials mainly focuses on four aspects: (1) The fabrication of negative Poisson's ratio materials through the stacking of specific angle-ply laminates; (2) producing negative Poisson's ratio composite by introducing negative Poisson's ratio inclusions or base materials; (3) studying and evaluating the properties of negative Poisson's ratio composite; (4) manufacturing processes of negative Poisson's ratio composite. Zorzetto et al. generated a new type of architectural composite material by combining two cellular structures with significantly different Poisson's ratios. They found that the addition of a small amount of concave negative Poisson's ratio material (about 12%) can significantly increase its stiffness (300%) without changing the overall relative density. Zahra et al. effectively improved the delamination and brittleness of cement composite by adding negative Poisson's ratio materials. Hu et al. prepared a novel elastic hydrogel-negative Poisson's ratio composite material by combining hydrogels and negative Poisson's ratio materials. They found that the hydrogel can temporarily repair fracture zones and cracks in the base material.

4.2 Fabrication of Porous Negative Poisson's Ratio Materials

4.2.1 Fabrication of Negative Poisson's Ratio Foam

In 1987, Lakes et al. fully compressed common foam material from three vertical directions into a special mold. The mold is heated to 163–171 °C (slightly above the softening point of the foam) and maintained at that temperature for some time. Remove foam after natural cooling. The isotropic Poisson's ratio of foams is −0.7. This was the first time that negative Poisson's ratio material produced by humans. Since then, the fabrication of materials with negative Poisson's ratio has been reported frequently [1].

4.2.2 *Fabrication of Honeycomb Structure with Negative Poisson's Ratio*

4.2.2.1 3D Printing

3D printing, also known as "additive manufacturing technology", originated in the late 1970s to the late 1980s. During its development, 3D printing was also known as "material accumulation manufacturing", "rapid prototyping", "layered manufacturing" and "solid freeform manufacturing" among others. The materials used in 3D printing mainly include engineering plastics, photosensitive resins, rubber materials, metal materials, and ceramic materials. In addition, cell biomaterials and food materials can also be used for 3D printing. There are various types of 3D printing manufacturing technologies, including Fused deposition modeling Rapid prototyping, Stereolithography, selective laser sintering, laminated object manufacturing, three-dimensional inkjet printing, etc. 3D printing only places materials where they are needed, which not only saves materials but also allows for more complex structures to be created [2].

The structure of a material determines its performance. For example, composite with discontinuous fiber structures exhibit significant fracture toughness, while materials with micro-truss structures maintain high strength even at low densities. Therefore, researchers strive to manufacture materials with superior properties through structural design. Due to the limitations of traditional fabrication methods in customizing material structures, complex geometries have only been achievable through 3D printing. In recent years, research on 3D printing inks and printing techniques has also played a significant role in producing materials with unique structures and properties.

The use of 3D printing to design specific structures can give materials unique properties such as negative Poisson's ratio or bistable state, improve the properties of materials in strength-density ratio, fracture toughness and other aspects, which has attracted the attention of many researchers [3].

Materials with negative Poisson's ratios produced through 3D printing are often porous. Clausen et al. designed a series of structures with adjustable Poisson's ratios using topology optimization programming (as shown in Fig. 4.1), and validated their properties by 3D printing. These materials exhibited Poisson's ratios ranging from −0.8 to 0.8, remaining constant within a deformation range of about 20%. Wang et al. utilized dual-material printing techniques to create structures with elastic connection points and rigid beams. This design prevented buckling of beams or walls during deformation, enabling the material to maintain a stable negative Poisson's ratio even under large strains. Li et al. printed a novel cell structure consisting of a hollow skeletal structure with concave angles. By using materials with different Poisson's ratios and varying geometric variables, cell structures with different Poisson's ratios were obtained. This structure's Poisson's ratio could be varied from positive to negative.

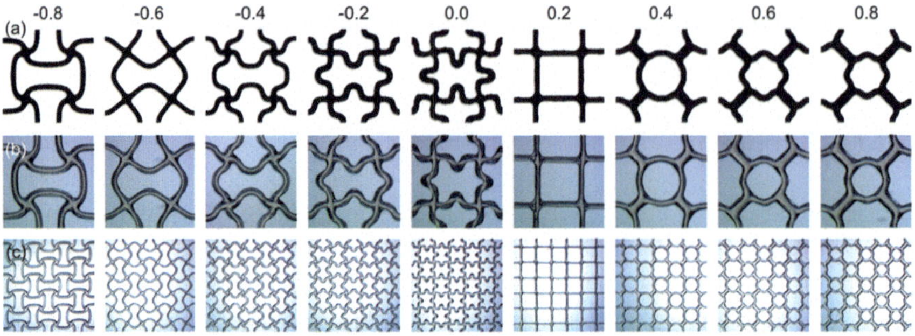

Fig. 4.1 Poisson's ratio controllable structure

The use of 3D printing enables the fabrication of materials with more complex structures, which can enhance certain properties and even impart unique characteristics. Therefore, 3D printing technology will become an important option for material modification. Further developments in 3D printing for material structure design can be focused on the following aspects: ① Improve the properties of existing materials, develop new printable materials; ② Develop the printing technology to improve the ability and accuracy of 3D printing to make different size structures, especially the ability to accurately make small size structures, as well as the ability of multi-material printing; ③ To increase the research on the structure and function of materials, provide rich design ideas for 3D printing materials.

Due to the small size and high requirement for detail representation of star-shaped and arrow-shaped honeycomb structure models, traditional mechanical manufacturing processes such as welding and joining cannot meet the experimental requirements. Therefore, in this paper, 3D printing rapid prototyping equipment will be used to select suitable materials for the fabrication of experimental models that meet the mechanical testing requirements of negative Poisson's ratio honeycomb structures. The negative Poisson's ratio honeycomb structure specimens will undergo relevant mechanical experiments [4].

3D printing technology is a rapidly developing cutting-edge emerging technology and is considered one of the important symbols of the third industrial revolution. Traditional manufacturing methods usually involve cutting, grinding, melting, and other methods to obtain the components of a structure from raw materials, and then synthesize the required products through techniques such as joining and welding. However, 3D printing technology is completely different and belongs to the category of Rapid Prototyping technology. Its basic principle is to slice a 3D model designed by software such as Solidworks and ProE into an infinite number of cross-sectional layers according to a plane coordinate system through scanning and other methods, and then print and stack them layer by layer to restore a solid 3D model. Compared with traditional manufacturing methods, 3D printing has many obvious advantages:

1. 3D printing technology does not need mold and manufacturing, can directly generate various shapes of objects from the computer design software, shorten

the production cycle of product development and preparation, improve work efficiency.
2. Products made by 3D printing technology are made by layering printing, and the connection performance and stability between structures are much higher than that of traditional manufacturing processes such as splicing and welding.
3. 3D printing mode changes the working mode between manufacturers and consumers. Consumers can participate in the process of product design, fully reflect the intentions of consumers, and meet the demand for structural performance.

In 3D printing, based on the purpose of model fabrication, there are generally two types of models according to the requirements: appearance verification model and structure verification model:

1. Appearance verification model: Engineers design and make hand models for product appearance verification or directly use models with high appearance requirements. Appearance verification model is visible and touchable, it can be very intuitive to reflect the designer's creativity in the form of real objects, avoid the shortcomings of design intention and production process error caused by the error of finished products. Appearance verification model making is essential in the process of new product development and product appearance review.
2. Structural verification model: In the process of product design from design scheme to mass production, it is generally necessary to make molds. The cost of mold manufacturing is very high, and the relatively large mold is worth hundreds of thousands or even millions. If the structure is unreasonable or other problems are found in the process of mold opening, the loss can be imagined. Therefore, making the structural verification model can avoid this loss and reduce the risk of mold opening.

In the process of creating models for product design, education, scientific research, medicine, and other fields, it is important to choose the appropriate materials based on the intended use of the model. Factors such as mechanical properties, chemical stability, strength, toughness, detail representation ability, surface smoothness, and post-processing rework ability are all important considerations when selecting materials for model fabrication. Common 3D printing materials include photosensitive resin (DSM Somos), nylon (PA12), glass-fiber nylon, and cobalt-chromium alloy.

Photosensitive resin is a low-viscosity photosensitive polymer used to manufacture robust, precise, and waterproof components. Parts made with white photosensitive resin are opaque and have properties similar to engineering plastics such as ABS and PC. It is generally used for achieving smooth surfaces and excellent detail representation, and is suitable for products with lower requirements for strength and temperature resistance.

Nylon (PA) possesses excellent mechanical properties, wear resistance, insulation, chemical stability, and solvent resistance. However, it has a certain level of moisture absorption, which significantly reduces its mechanical and insulation properties. Parts made from nylon have high strength and a certain level of flexibility, allowing them to withstand impact and resist pressure when bent. The surface of nylon has a slightly rough, powdery texture. It is commonly used for structural validation or for models with specific requirements for strength and toughness.

Nylon glass fiber is made of nylon (PA) adding 30% glass fiber, compared with nylon (PA), its mechanical properties, dimensional stability, heat resistance, aging resistance has been significantly improved, fatigue strength is enhanced 2.5 times. The molding process of glass fiber reinforced PA is roughly the same as that of unreinforced PA. The material is the combination of nylon material and glass fiber, has the performance of both, belongs to the reinforced version of nylon material. Compared with nylon material, its strength, toughness is greatly improved. It can also be used for structural verification, and is often used for complex parts or hollow modeling.

To better demonstrate the negative Poisson's ratio performance of star and arrow-shaped honeycomb structures, this paper utilized photosensitive resin material for fabricating specimens for tensile testing. This material exhibits excellent detail representation capabilities and effectively showcases the negative Poisson's ratio expansion properties of honeycomb structures. Star and arrow-shaped honeycomb structures were designed using Solidworks software, and the corresponding experimental specimens were rapidly prototyped using 3D printing equipment, as shown in Fig. 4.2. Tensile testing was conducted on the photosensitive resin material specimens, yielding a Young's modulus E of 2560 MPa.

Fig. 4.2 Negative Poisson's ratio honeycomb structure fabricated by 3D printing rapid prototyping equipment

4.2.2.2 Mold Making

This study utilized aramid paper combined with epoxy resin to prepare honeycomb structures with negative Poisson's ratio.

(1) Mold making

In order to reduce the difficulty of honeycomb preparation, the Poisson's ratio of the honeycomb structure to be made was set as -1, and the height was set as 1/2 of the thickness of the wing beam of the electric fiber composites intelligent drive structure in Chap. 3, that is, 12 mm. The function of the mold is to facilitate solidification of honeycomb. Because the material of honeycomb production is aramid paper and epoxy resin, it can only choose the way of winding the epoxy resin infiltrated by aramid paper onto the mold for curing molding, which requires the mold to have the shape of honeycomb. Considering the error caused by human labor during the test, $\theta = 28°$ is set during the mold production. Mold height is 15 mm, honeycomb cell $h = 10$ mm, $l = 5$ mm. A laser cutter is used to cut the plexiglass into equilateral trapezoid, and then glue the small pieces of the cut plexiglass equilateral trapezoid to make mold bumps, as shown in Fig. 4.3. In order to facilitate the production of post-release, the mold convex blocks are pasted with release cloth. A laser cutting machine was used to cut 2 mm thick plexiglass plate into rectangular strips with equilateral trapezoid evenly spaced 0.5 mm as mold substrate. Mold substrate was affixed with a stripping cloth, and the stripping cloth on the equilateral trapezoid hole was holed out. Finally, the honeycomb mold convex block with the stripping cloth is glued to the honeycomb mold base plate. Figure 4.4 shows the mold produced [5] (Fig. 4.5).

(2) Curing molding of honeycomb

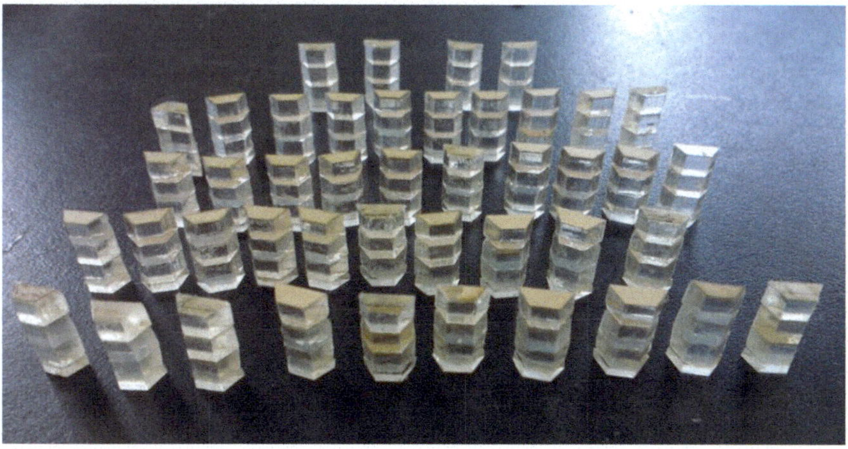

Fig. 4.3 Honeycomb mold bumps with negative Poisson's ratio

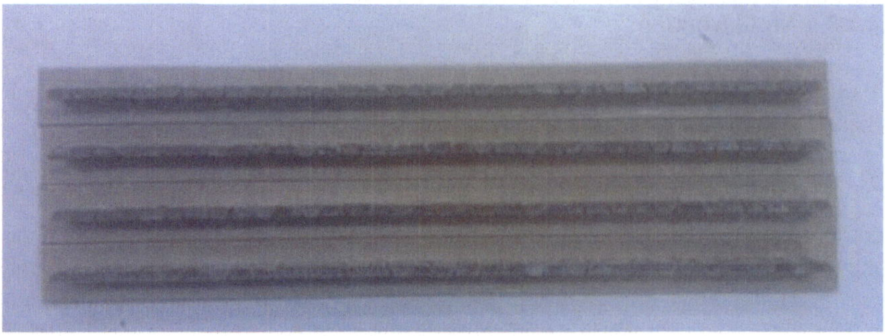

Fig. 4.4 Honeycomb mold with negative Poisson's ratio

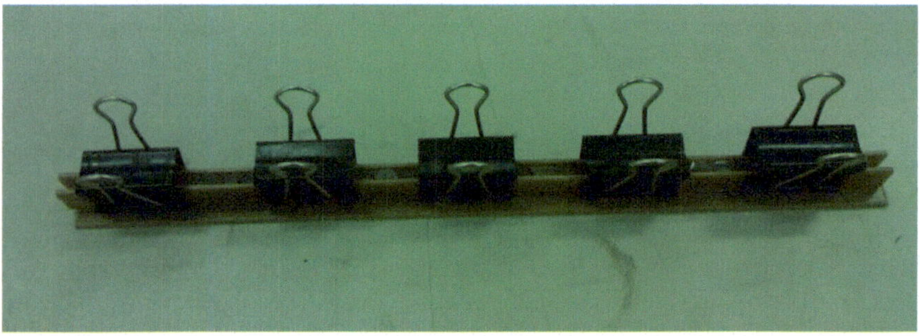

Fig. 4.5 Honeycomb solidification molding

Cut the aramid paper into a 12 mm wide strip, and configure adhesive which epoxy resin: curing agent: diluent solvent is 1:0.5:2. The adhesive is evenly spread on the aramid paper strip by the coater, and then another aramid paper is covered on the aramid paper coated with epoxy resin. The paper is put on the coater for compaction and extrusion of the bubbles inside, so as to ensure that the two aramid paper are completely covered. The epoxy resin is applied in the middle of the two aramid paper to reduce the difficulty of stripping honeycomb paperboard after curing and forming. Then wrap the prepared Aramid strip around the mold made in step (a), clamp it with two long strips attached with the release cloth, and fix it with a long tail clip, as shown in the figure. The assembly was placed in a high-temperature drying oven at 60 °C for one hour to cure. After demolding, the resulting honeycomb paper strips were obtained, as shown in Fig. 4.6.

(3) Bonding molding of honeycomb

The honeycomb paper strips were cured and molded multiple times, and the cured honeycomb paper strips were bonded together using a fast-drying adhesive, specifically 502 glues. The resulting honeycomb structure with a negative Poisson's ratio is shown in Fig. 4.7.

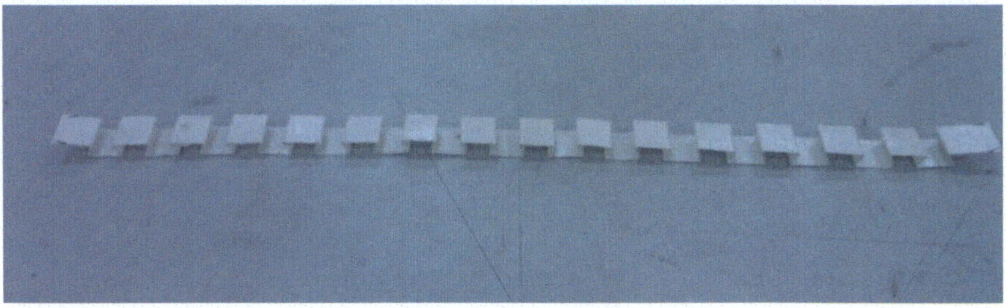

Fig. 4.6 Honeycomb paper strip after solidification molding

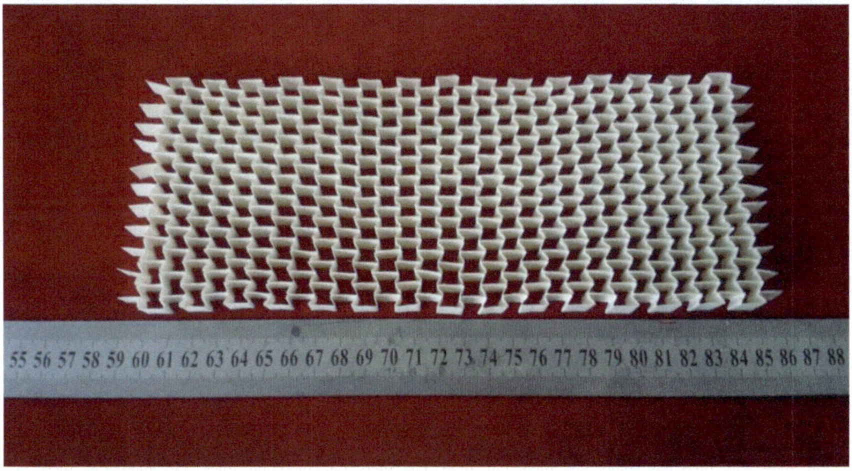

Fig. 4.7 Negative Poisson's ratio honeycomb structure obtained from the test

4.2.2.3 Other Manufacturing Methods

Wei Plateau et al. created a concave honeycomb structure on ethylene–vinyl acetate copolymer through mechanical processing, resulting in an auxetic porous material. In 1999, Evans et al. utilized femtosecond laser cutting technology to cut a concave honeycomb structure with an aperture of approximately 1mm on a polyester sheet, producing a stretchable filtration mesh with a Poisson's ratio of −2 [6].

The preparation method of concave hexagon negative Poisson's ratio honeycomb, including the following steps: Step 1: the raw material base material is coated, laminated, and then hot pressed to obtain honeycomb stack; Next, the honeycomb stack is stretched and shaped to obtain the overstretched white honeycomb the cross section shape of the hole of the overstretched white honeycomb is rectangular; The second step: the overstretched white honeycomb prepared in the first step is soaked for the first time, and then the heat treatment is carried out to make the soaked glue solution semi-cured; The third step: on the construction frame, the hole of the stretching

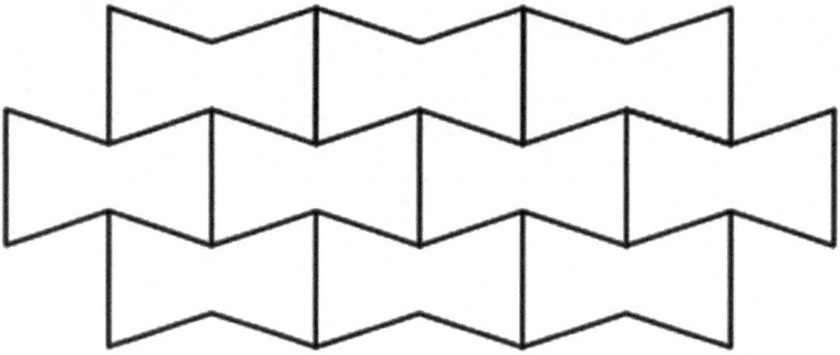

Fig. 4.8 The structure of concave hexagon negative Poisson's ratio honeycomb

white honeycomb is compressed, the compression is the middle of the hole long side, the hole is bent to the middle of the long side; The fourth step: the compressed honeycomb is impregnated with glue and fixed to obtain hexagon honeycomb with negative Poisson's ratio (Fig. 4.8).

4.3 Fabrication of Polymer Materials with Negative Poisson's Ratio

In 1992, Alderson and Evans prepared negative Poisson's ratio material with negative Poisson's ratio of −1.24 by pressing ordinary polyethylene, temperature sintering and hot extrusion. In the special extrusion equipment shown in Fig. 4.9, the polyethylene powder was pressed into a pressure plastic rod with a diameter of 15 mm and a length of 80 mm. The pressing speed was 20 mm/min, and the pressure was controlled by 0.04 GPa for 20 min. The die is heat-treated at high temperature that 130–200 higher than the melting point of polyethylene. C temperature, then the die is hot extruded. It has also been reported that a press mold filled with polymer powder is heated to 110–125 °C within 3–10 min, and then pressed into a specific press tank at a rate of 140 mm/min for a constant pressure of 0.04 GPa for 10–20 min. The die is then transferred to the metal impression and cooled naturally, then reheated to 160 °C. After constant temperature for 20 min, the particles were quickly extruded at a high speed of 500 mm/min. The upper diameter was 7–7.5 mm, the lower diameter was 15 mm, and the length was 3.3. It was confirmed that the treated polymer had a negative Poisson's ratio.

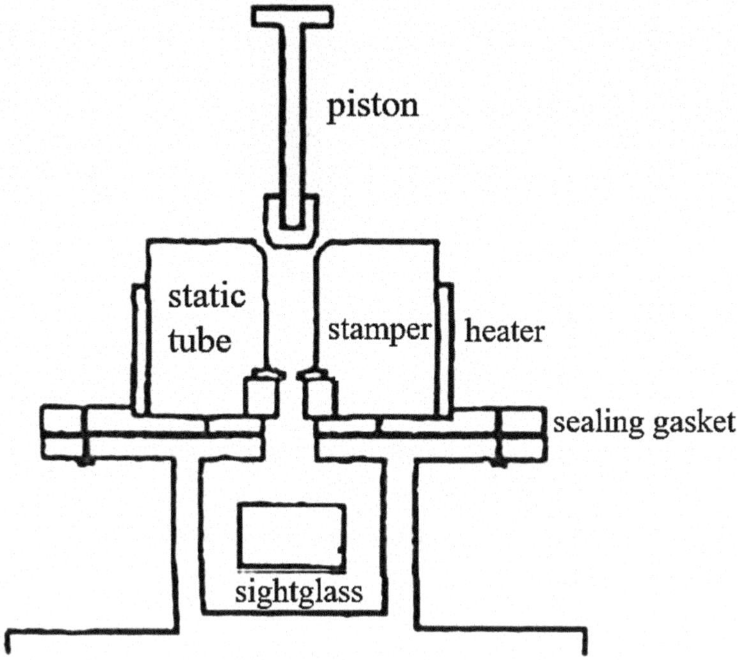

Fig. 4.9 Schematic diagram of extrusion equipment used by Alderson and Evans

4.4 Fabrication of Negative Poisson's Ratio Composites

In 2005, K. L. Alderson and V. R. Simkins fabricated a negative Poisson's ratio fiber, as shown in Fig. 4.10b, using the apparatus depicted in Fig. 4.10a. Powdered polymer was introduced into a column 2 equipped with a screw 3 via a funnel 1. The column 2 consisted of three sections: feeding 4, compression 5, and measurement 6. Connected to the end of the column 2 were an adapter region 7 and a die 8 with a 40-mesh spinneret. The extruded polymer filament passed through a cooling roller 9 with a compression roller 10 and an air knife 11, then was guided by guide shafts 12–14 and a fence 15, and finally entered a high-speed rotating roller 16.

In 2007, K. L. Alderson et al. once again produced negative Poisson's ratio fibers using the "fusion-spinning" technique, and investigated the effects of temperature, spiral speed, extrusion speed, and impression size on the material's Poisson's ratio.

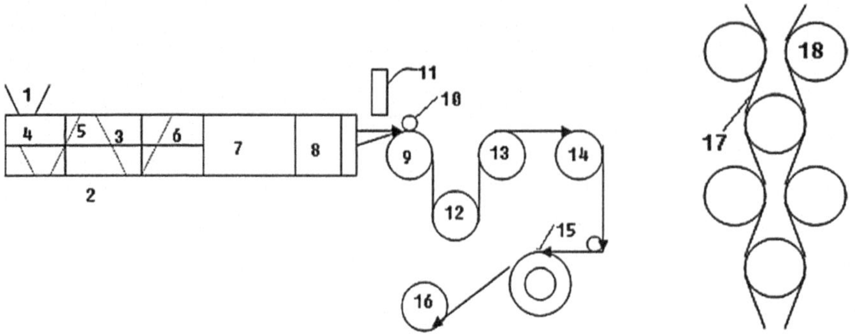

(a) device (b) Negative Poisson's ratio fibers

Fig. 4.10 Device diagram for producing negative Poisson's ratio fibers

4.5 Negative Poisson's Ratio Measurement of Poisson's Ratio of Materials

After the fabrication of negative Poisson's ratio materials, it is necessary to measure their negative Poisson's ratio parameters in order to gain a detailed understanding of their properties and perform relevant verification.

Elastic modulus and Poisson's ratio are two important parameters that characterize the mechanical behavior of materials, and the key issue in all studies on Poisson's ratio is the accurate measurement of Poisson's ratio. In order to verify the negative Poisson's ratio effect of star and arrow-shaped honeycomb structures, the lateral displacement variables generated during tensile testing of the experimental components need to be measured. There are many methods for Poisson's ratio testing, which can be classified into mechanical methods, acoustic methods, optical methods, etc., depending on the different basic principles used in the Poisson's ratio testing process.

Mechanical methods for measuring Poisson's ratio of materials generally belong to contact measurement. The American Society for Testing and Materials (ASTM) requires the use of two pairs of extensometers to measure the transverse and longitudinal strains of the structure. The national standard requires the value to be indirectly calculated by directly measuring the E and G values through formulae. E can be measured by uniaxial tensile, compressive, or bending tests, while G can be measured by torsion tests or simple shear tests. Acoustic methods are completely non-destructive and use a laser beam to directly excite surface waves on the sample surface. Optical methods belong to non-contact measurement, including optical interferometry, photoconductive holography and full-field strain measurement based on digital image correlation method (DIC). In addition, the commonly used electrical measurement method using resistance strain gauges is based on the principle of converting the change in resistance value of the strain gauge during the deformation

Table 4.1 Comparison between non-contact displacement measurement system and traditional measurement instruments

	Non-contact displacement measurement system	Laser displacement meter	External strain gauge
Operation mode	Non-contact	Non-contact	Contact
Measuring range	Large range	Small range	Limited
Measuring environment	Applied widely	Conventional conditions	Conventional conditions
Measuring direction	Multipoint 3D measurement	Single direction	Single direction
Measurable parameters	Displacement, speed, acceleration, strain.et	Displacement	Displacement

of the specimen into a change in strain value to obtain the Poisson's ratio of the material.

Different Poisson's ratio test methods have different advantages and disadvantages. Mechanical measurement method is simple and convenient to operate in the conventional measurement environment, and the displacement measurement is intuitive. However, due to the large dead weight and holding force of extensometer, additional deformation of soft pattern is caused. The sensitivity of the electric measurement method is high, but the resistance plate affixed to the specimen itself has certain requirements, and the measurement range is limited. While the optical method has higher test accuracy, requires image capture and processing equipment, so costs higher. The non-contact full-field strain measurement system based on digital image correlation method (DIC) has many advantages, such as full-field measurement, non-contact, simple optical path, wide application range and so on.

In conclusion, considering the limitations of traditional sensors as shown in Table 4.1, this study utilizes a non-contact full-field strain measurement system, VIC-3D, to achieve this objective.

4.5.1 Digital Image Method

The non-contact full-field strain measurement system, VIC-3D, based on digital image correlation (DIC), enables the calculation of surface node displacements of an object. The basic method involves capturing images of the object before and after deformation using an image acquisition system. The obtained images are then subjected to computational analysis to obtain strain data for the nodes. Horizontal and vertical displacements are calculated, and subsequently, the Poisson's ratio can be determined. The relevant equipment is illustrated in Fig. 4.11.

Firstly, the camera distance, angle, focal length, and aperture are adjusted according to the size of the object being measured. A small aperture and long focal

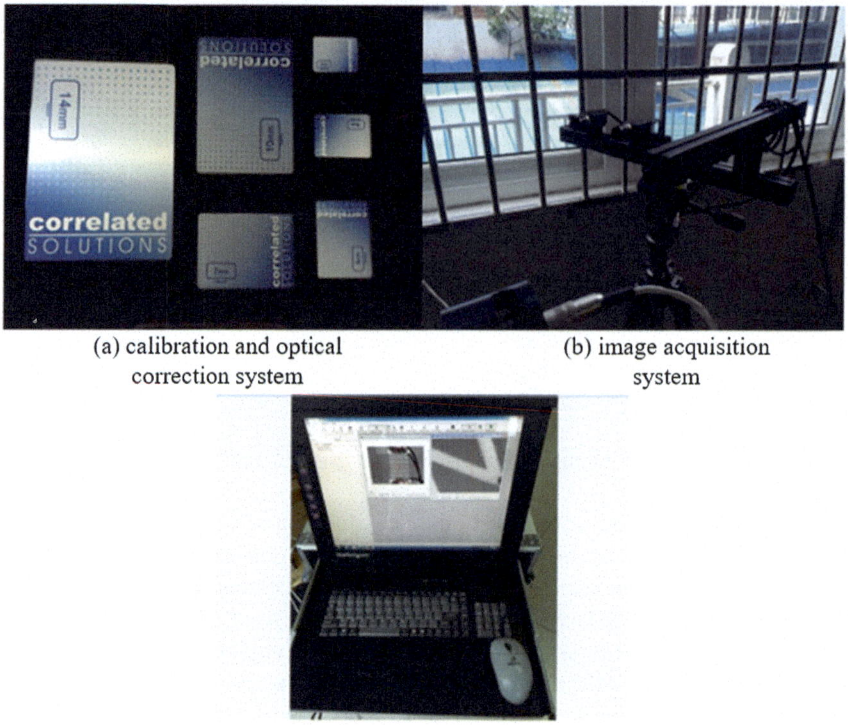

(a) calibration and optical correction system

(b) image acquisition system

(c) control and computing systems

Fig. 4.11 VIC-3D composition system

length are used in sufficient external lighting conditions, and the camera is positioned as close to the measurement midpoint as possible to avoid the measurement error caused by the distortion of imaging. The image acquisition system is calibrated and optically corrected using calibration and optical correction plates to ensure parallelism with the experimental plane and clear imaging, thereby identifying the position information and parameters of the measurement object and meeting the experimental requirements. Data points requiring measurement are selected on the surface of the experimental component and marked using special tools, as shown in Fig. 4.12.

Adjust the focal length of the lens of the image collector, select the area captured in the experiment, and set relevant parameters for the DIC digital image correlation analysis software. The non-contact displacement measurement system could set the acquisition frequency and camera exposure time as required, obtain the measurement data through the control and calculation system, and analyze the data of the selected nodes.

In this paper, tensile tests were conducted on star-shaped and arrow-shaped negative Poisson's ratio honeycomb structures, following the requirements of the national

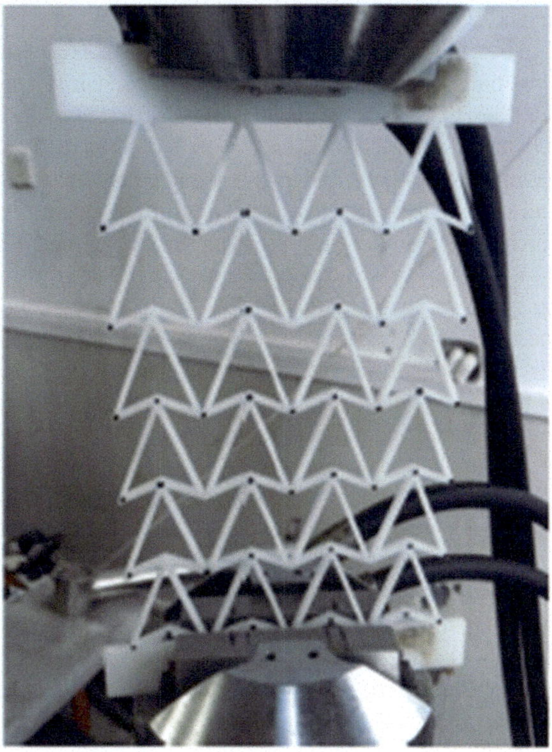

Fig. 4.12 Data measurement point selection

standard GB/T1452-1987, "Test Method for Tensile Strength of Non-metallic Sandwich Structures." The honeycomb structures tested had a side length or diameter of 60 mm, or at least included 4 complete cells. The MTS universal testing machine was used to perform the tensile tests on the honeycomb specimens.

4.5.2 Tensile Test Stand

The honeycomb tensile test was carried out on the tensile tester where Poisson's ratio was measured. Make four clamping pieces and fix the clamping pieces at both ends of the honeycomb with a long tail clip. Ensure that the clamping pieces are as perpendicular to the stretching direction of the honeycomb as possible. Use organic glass plate to make a honeycomb support platform. Note that the platform should be as smooth as possible to ensure that the honeycomb will not be affected by the friction force of the support platform during the stretching process. Lay a piece of coordinate paper on the support platform, remove the tail handle of the long tail clip on the upper surface of the honeycomb, put the tail handle of the long tail clip under the movable chuck of the tensile test platform, tighten the screws. Fix the honeycomb and the coordinate paper together, and take this end as the fixed end, the other end as the stretching end. Try to ensure that the long axis of the honeycomb is parallel to the stretching direction of the stretching test bench When tightening the screw. In order to ensure that the stretching direction is parallel to the axis direction of the stretching test stand, a nylon rope is used to pass through the long tail of the stretching end and clamp the tail shank, and then tied and wrapped around the bolt fixed to the movable end of the stretching test stand. Before stretching, ensure that the honeycomb is spread freely on the supporting platform, and mark the honeycomb cross section 1, 2, 3, 4, 5, 6, 7, 8 and 9 and the position of the gripper on the graph paper under the honeycomb. Then turn the handle of the active end of the stretching test stand, so that the honeycomb is gradually in the stretched state. After the honeycomb extends for a distance, stop turning the handle, and mark the position of the gripper and the position of the honeycomb cross section 1, 2, 3, 4, 5, 6, 7, 8, 9 on the coordinate paper under the honeycomb again. Note that because the thickness of the Kevlar paper-epoxy film used to make the honeycomb is very small, the elastic modulus of the honeycomb in the tensile direction is also very small. As a result, if the stretching distance is too large, the θ of the honeycomb cell will become smaller, which will affect the test results. However, if the stretching length is too small, it will cause trouble in the measurement of the test data. So that the final results will still produce a large error, so the tensile length of the laboratory should be guaranteed within a certain range, which will not affect the θ, but also easier to measure the data. Poisson's ratio of the cell is calculated by using the displacement of the gripper and the average elongation of nine cross sections of the cell. The experiment was repeated three times, and the average value of each calculated Poisson's ratio was taken as the cellular Poisson's ratio (Figs. 4.13 and 4.14).

The strain in X and Y directions is calculated as follows:

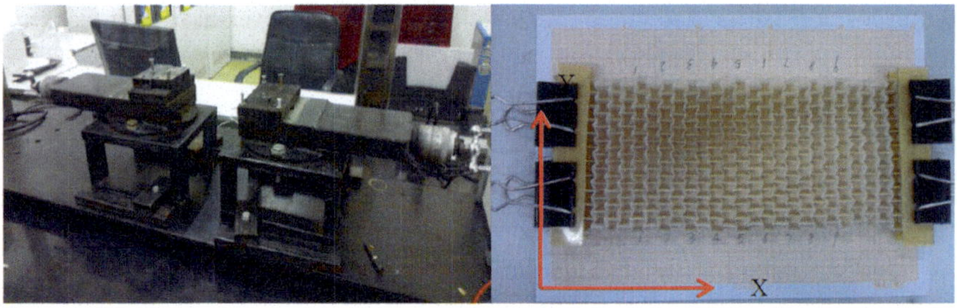

Fig. 4.13 Honeycomb structures Poisson's ratio measurement scheme

Fig. 4.14 The negative Poisson's ratio honeycomb fixed on a tensile test bench

$$\varepsilon_x = \frac{1}{2}\left(\frac{\Delta l_{上}}{l_{下}} + \frac{\Delta l_{下}}{l_{下}}\right)$$
$$\varepsilon_y = \frac{1}{9}\left(\frac{\Delta h_1}{h_1} + \frac{\Delta h_2}{h_2} + \frac{\Delta h_3}{h_3} + \frac{\Delta h_4}{h_4} + \frac{\Delta h_5}{h_5} + \frac{\Delta h_6}{h_6} + \frac{\Delta h_7}{h_7} + \frac{\Delta h_8}{h_8} + \frac{\Delta h_9}{h_9}\right) \quad (4.1)$$
$$v_{xy} = -\frac{\varepsilon_y}{\varepsilon_x}$$

The Poisson's ratio of honeycomb is obtained by eliminating the two groups of values with large errors.

4.6 Summary of This Chapter

The chapter provides an overview of various manufacturing methods for negative Poisson's ratio structures. The first section discusses the research achievements and recent advancements in negative Poisson's ratio materials, highlighting different types of such materials. The second section provides a detailed account of the manufacturing process for porous negative Poisson's ratio materials, including the creation of such structures using foam materials, as well as the techniques of 3D printing and mold manufacturing for honeycomb-like negative Poisson's ratio structures. The third and fourth sections provide detailed descriptions of the manufacturing processes for negative Poisson's ratio polymers and composite. The fifth section focuses on the measurement methods for the Poisson's ratio of negative Poisson's ratio materials, specifically digital image correlation and tensile testing. By presenting these various manufacturing methods for negative Poisson's ratio materials, the chapter serves as a reference for proposing new manufacturing approaches.

References

1. Huang Zhi, Wang Wanlu, Liao Kejun, et al. Research progress on materials with negative Poisson's ratio [J]. Materials Review, 2002(10):49–51
2. Zhang Hao, Huang Guangsu, Wu Jinrong. 3D printing of specific structure materials [J]. Plastics Industry, 2016.
3. Ma Jie. Design and mechanical properties research of novel structures with negative Poisson's ratio materials [D]. Northeast University, 2018.
4. Zhang Yifan. Numerical analysis and experimental research on two types of honeycomb structures with negative Poisson's ratio [D]. Jinan University, 2015.
5. Huang Jian. Research on twistable wing structures based on piezoelectric fiber composite materials [D]. Harbin Institute of Technology, 2013.
6. Li Jing, Wang Yu, Wu Hongmei, et al. Research progress on materials with negative Poisson's ratio [J]. Jiangxi Chemical Industry, 2016(01):69–73.

Chapter 5
Engineering Applications of Negative Poisson's Ratio Structures

5.1 Introduction

The special mechanical properties of negative Poisson's ratio materials and their unique causal mechanisms provide novel approaches and methods for the application of materials with special mechanical properties, rendering them a wide range of potential applications [1]. According to the summary of the research on their properties, negative Poisson's ratio materials can be mainly used in the following aspects: (1) using the material compression-shrinkage properties, they can be used to make bullet-proof equipment, etc.; (2) using their properties in bending, they can be used for molding and forming panels for aircraft and automobiles; (3) using their compression properties, they can be used to make structures that are more easily embedded in the parent body; (4) use its special energy-absorbing properties to make energy-absorbing equipment with excellent performance; (5) make highway protection devices and high-performance protective equipment to resist external impact; (6) use it in the medical field to make artificial organs with high load resistance; (7) use it to make vibration-absorbing structures with excellent performance.

Despite the belief among scholars regarding the extensive applications of negative Poisson's ratio materials in various fields, most of them are still at the hypothetical and theoretical level, and further comprehensive studies are required to effectively apply them to engineering practice, while some scholars have also done in-depth research and exploration on the application of negative Poisson's ratio materials in industry. Garber et al. applied thermally expanded graphite with negative Poisson's ratio characteristics in aircraft to enhance its thermal protection capability [2]. Jones incorporated the negative Poisson's ratio foam into headphones to make the headphones better cover the irregular shape of the ear, and the headphones have better comfort and less air leakage, etc. [3]. Mitsubishi Heavy Industries utilized negative Poisson's ratio materials in bullets to minimize the expansion of bullets in the barrel, improving bullet movement by achieving a zero equivalent Poisson's ratio. Chang Qi et al. introduced the negative Poisson's ratio structure into the seat belt,

G. Zhou et al., *Design, Manufacturing, and Application of Negative Poisson's Ratio Structure*, https://doi.org/10.1007/978-981-96-1937-5_5

proposed the concept of negative Poisson's ratio seat belt, and studied the influence of Poisson's ratio value and friction factor on the protection efficiency of the seat belt. They also explored and verified the influence of negative Poisson's ratio material performance through computer simulations. Furthermore, Yan Fangfang and team proposed the potential application of negative Poisson's ratio materials in the variable wing design of unmanned aerial vehicles (UAVs) as a means to enhance their takeoff and landing performance. They substantiated the structural performance through in-plane tensile and compression tests. A new type of marine honeycomb vibration isolator with negative Poisson's ratio effect was proposed by Zhang, Teruelin et al. and the vibration damping effect of negative Poisson's ratio materials was discussed by computer simulation [4]. The studies conducted by researchers on the application of negative Poisson's ratio structures have significantly contributed to the widespread utilization of negative Poisson's ratio materials. However, it is essential to acknowledge that while negative Poisson's ratio materials possess special mechanical properties, they also have some drawbacks compared to traditional materials. These drawbacks include lower hardness, reduced stiffness, and lower strength. Therefore, when applying negative Poisson's ratio materials in industrial design and manufacturing, it is crucial to investigate methods of effectively leveraging the performance advantages of these materials while mitigating their disadvantages. This may involve finding ways to optimize the special mechanical properties of negative Poisson's ratio materials while addressing their limitations through appropriate strategies and material modifications.

5.2 Application of Negative Poisson's Ratio Structure in Energy Absorption

In summary, the unique mechanical properties and lightweight structure of negative Poisson's ratio materials indicate their substantial potential for various applications in the automotive industry. Therefore, conducting research on the investigation methods, distinctive mechanical properties, and microstructure design of negative Poisson's ratio materials hold significant practical importance. Such studies can provide a theoretical foundation for the widespread adoption of negative Poisson's ratio structures and materials within the automotive industry. At the same time, the application of negative Poisson's ratio structure in the design of automobile structure is bound to help further improve the safety performance and lightweight design of automobiles, which is of great significance to the development of automobile design and manufacturing.

At present, research on negative Poisson's ratio structures or materials primarily focuses on foam and composite, which are easy to manufacture, but these materials lack periodicity in their microstructure, and the size of the microstructure in the material is small. Consequently, it is challenging to design the microstructure based on the material's performance requirements during application, leading to variations in mechanical properties. This poses difficulties in the design and manufacturing processes, making it challenging to obtain satisfactory materials. With the continuous research on the effect of negative Poisson's ratio and the rapid progress of industrial processing technology, people gradually extend from the microscopic material field to the macroscopic structure field. Compared with negative Poisson's ratio materials, negative Poisson's ratio structures offer more control over the size and shape of microstructures, making them more suitable for wide-ranging engineering applications. The research on negative Poisson's ratio structures began with two-dimensional structures. Two-dimensional periodic structures like the H-shaped lattice and star-shaped lattice exhibit periodicity and are easily controllable. Then, when it is applied, the negative Poisson's ratio characteristics can only be reflected in a specific axial compression or tension, so its application is greatly limited, and a more effective negative Poisson's ratio structure needs to be found. The two-dimensional double-arrow negative Poisson's ratio structure was initially obtained by Ulrik and Ole using a topology optimization method. Through setting material property requirements during the optimization process, they achieved the two-dimensional double-arrow negative Poisson's ratio structure [5]. Compared with the general negative Poisson's ratio structure, the two-dimensional double-arrow structure exhibits a higher negative Poisson's ratio effect, thereby enhancing the performance of parts manufactured using this structure. However, its application is somewhat limited due to its two-dimensional nature. With the rapid development of 3D printing and other manufacturing levels, ZhengDong Ma's research team has extended the two-dimensional double-arrow negative Poisson's ratio structure to three dimensions. With the three-dimensional structure and controllable microstructure rules, this greatly expands the application scope of the double-arrow negative Poisson's ratio structure and makes the double-arrow negative Poisson's ratio structure have a wider application prospect. Moreover, compared with the two-dimensional double-arrow negative Poisson's ratio structure, the three-dimensional double-arrow negative Poisson's ratio structure has more obvious negative Poisson's ratio characteristics. In summary, the remarkable negative Poisson's ratio characteristics, easy extension to three dimensions, controllable structure rules, and typicality make the double-arrow negative Poisson's ratio structure a crucial research subject when studying negative Poisson's ratio structures. The simulation analysis and optimization method used in this paper can also be used in the study of other negative Poisson's ratio structures and materials.

5.2.1 Application of Negative Poisson's Ratio Structure in Automobile Energy Absorbing Box

With the improvement of people's living standards, the annual production and ownership of cars in China are increasing rapidly. Due to the safety awareness and traffic facilities, the frequency of automobile traffic safety accidents in China is at a high level, which brings serious threats to people's lives and property safety. Therefore, improving the crashworthiness of automobiles to reduce the accident casualty rate is a key research area in automobile design. The bumper system plays a vital role in improving a vehicle's crashworthiness. During a collision, the bumper system absorbs energy through its deformation and distributes the collision force to other components, thus minimizing the risk of injury to occupants [6]. The energy-absorbing box is the main energy-absorbing component of the bumper system, and the performance of the box has an important impact on the energy absorption of the bumper system and the crashworthiness of the vehicle [7].

In order to improve the energy absorption performance of the energy-absorbing box, scholars at home and abroad have conducted in-depth researches on the design methods, theories and applications of the energy-absorbing box. Alexander studied the computational model of energy absorption through the deformation of metal circular tubes, providing a theoretical foundation for the crashworthiness of thin-walled structures [8]; Lanzi investigated the energy absorption performance of energy-absorbing boxes under various working conditions and applied optimization algorithms to optimize the box structure's design, thereby improving its energy absorption performance [9]; Hosseinipour pointed out through his research that a reasonable increase in the induced groove could significantly improve the energy absorption characteristics of the box, thus improving the crashworthiness of the whole vehicle [9]. Hao, Liang et al. optimized the number, spacing and depth of the induced slots by establishing a multi-objective optimization model, which led to a large improvement in the performance of the energy-absorbing box structure [10]. Milin et al. analyzed and studied single- and multi-cavity aluminum alloy energy-absorbing boxes with different cross-sectional shapes to identify structures with optimal energy absorption characteristics [11]. Nagel et al. analyzed and studied the energy absorption performance of rectangular cross-sectional tapered tubes under dynamic loads and the effects of structural geometric parameters such as angle and thickness on the energy absorption performance of the structures [12]; Acar conducted an analysis on the effect of induced slots in circular cross-sectional tapered tubes, specifically examining their impact on the specific energy absorption and collision force efficiency of the structure [13]; Liu investigated the simplified models of ortho-hexagonal cross-section, rectangular cross-section and slotted thin-walled beam [14]; Li Bangguo et al. achieved a well-defined deformation mode and relatively smooth axial reaction force during the energy absorption process through rational structural design [15]. The above different studies on energy-absorbing boxes have greatly promoted the structural design level of energy-absorbing boxes, so that the boxes can absorb more energy under the same conditions or in shorter time, which significantly improves the crashworthiness level of the car. However, in order to further enhance the car's crashworthiness and better protect occupants, higher performance expectations are placed on the crash performance of energy-absorbing boxes. Traditional structural optimization design methods face limitations in terms of design space and materials, making it challenging to further improve the crash performance of energy-absorbing boxes.

Based on the special mechanical properties of the double-arrow negative Poisson's ratio structure, especially the energy absorption characteristics of the negative

Poisson's ratio structure under the action of impulsive load, this paper proposes a new type of structural energy-absorbing box for the first time—double-arrow negative Poisson's ratio structure energy-absorbing box. By forming a new double-arrow type negative Poisson's ratio energy-absorbing box with a common hollow thin-walled structure and a negative Poisson's ratio structure filled core, the energy-absorbing performance of the structure is improved while the deformation pattern of the box remains unchanged, so that it shows stronger energy absorption capability and thus improves the crashworthiness of the vehicle.

When a conventional material is subjected to an external force, compression occurs in the direction of the external force and the material flows in all directions perpendicular to the impact direction. Unlike conventional materials, the double-arrow negative Poisson's ratio structure is compressed in the direction of external force and also in the direction perpendicular to the external force, and the material flows from all around to the part where the compression occurs, and the local density of the material increases rapidly. In the design of crash energy absorbing structures of automobiles, it is difficult to continue to improve the energy absorbing properties of the structures by traditional structural optimization methods, so the application of negative Poisson's ratio structure in the design of energy absorbing structures of automobiles provides another effective way to improve the performance of energy absorbing boxes.

5.2.1.1 Establishment of Double-Arrow Type Negative Poisson's Ratio Energy Absorption Box Model

Due to the increasing attention to safety and stricter safety regulations, crashworthiness design has been the focus of attention during the development of automobiles, and the quality of such designs directly impacts the extent of vehicle damage and potential occupant injuries during collisions [16]. The energy-absorbing box is generally arranged in the front end of the longitudinal beam of the car, and is a thin-walled structure of the compartment type, whose main function is to reduce the damage to the front longitudinal beam and related main frame structure of the car by absorbing energy as much as possible during the low-speed collision of the car, and to reduce the repair cost of the car [17]. During the engineering design of energy-absorbing boxes, each component of the vehicle has different stiffness characteristics. If the collision force exceeds the limit, the connected parts may sustain damage, which necessitates strict control over the peak collision force endured by the energy-absorbing box structure. Increasing the length of the box allows for greater energy absorption, but practical constraints such as car shape and cost impose limitations on its overall length. Consequently, if the energy-absorbing box can absorb more energy per unit length, it can achieve the same energy absorption with a shorter length compared to other structures of equal length [18]. The advantages and disadvantages of the energy-absorbing box design largely affect the crashworthiness of the car, so it is especially important to design a reasonable energy-absorbing box structure [19]. During the design of the energy-absorbing box, the design goal should be to make the box absorb as much energy as possible by smoothly collapsing under the same crash conditions, and the collapse of the box during the crash should be smooth, regular and controllable, while the box should have a small compression distance and a smooth collision force peak that does not exceed the permitted force.

When evaluating automotive energy-absorbing boxes, several key evaluation criteria are commonly used, including total energy absorption, peak crash force, specific energy absorption, crash force efficiency, and average crash force [20]. The total energy absorption is the sum of energy absorbed by the energy-absorbing box through the deformation of the structure during the whole crash of the vehicle, which is reflected in the load curve as the area enclosed by the curve and the coordinate axis, and this index is used to characterize the energy absorption capacity of the energy-absorbing box. In the same collision conditions and limited deformation space, the energy-absorbing box should absorb as much energy as possible to reduce the damage to the main structure and occupants of the car. The specific energy absorption refers to the total energy absorbed per unit mass of the structure. Due to the limitation of the structure space arrangement and production cost, the specific energy absorption (*SEA*) of the structure is generally desired to be as large as possible.

$$SEA = \frac{E_{total}}{M_{total}} = \frac{\int_0^{\Delta} F(x)dx}{M} \tag{5.1}$$

where E_{total} indicates the total energy absorption, M_{total} is the total mass of the energy-absorbing box, and $F(x)$ is the function that indicates the crush force and the crush volume. The average collision force $F_{average}$ refers to the average value of the collision force during the whole collision process, as shown in Eq. (5.2), and this index reflects the level of average energy absorption of the energy-absorbing box structure to a certain extent.

$$F_{average} = \frac{E_{total}}{\delta} \tag{5.2}$$

where δ is the crush displacement. The collision force efficiency of the energy-absorbing box can be expressed as

$$CFE = \frac{F_{average}}{P_{\max}} \tag{5.3}$$

where P_{max} is the peak of the collision force, the peak of the collision force is the largest possible value of the collision force during the collision. Since the peak collision force directly affects the maximum acceleration, the peak collision force cannot be too large in the collision process, so as to ensure the crashworthiness of the vehicle and the safety of the members. The first peak occurs when the energy-absorbing box starts to buckle, which is mainly influenced by the elastic–plastic buckling strength of the structure. The second peak is when the energy-absorbing structure is fully compressed, and the collision force rises sharply because there is no deformable space. The appearance of the first peak in the collision process has an important influence on the structural failure, so the first peak is mainly analyzed in the performance study of the energy-absorbing box.

In order to improve the design level of the energy-absorbing box as much as possible, domestic and foreign researchers have done a lot of research on the design and improvement of the energy-absorbing box. These studies have significantly contributed to the understanding of the performance of the energy-absorbing box and the optimization of the structure shape and wall thickness to improve the energy absorption performance of the structure, which has greatly improved the design level of the energy-absorbing box and the crashworthiness of the vehicle. However, these studies are all aimed at the optimization of the shape of the general energy-absorbing box structure, and there are problems such as limited optimization space and potential. In this paper, we propose a thin-walled energy-absorbing box based on the double-arrow negative Poisson's ratio structure from the engineering reality and use the special energy absorption performance and lightweight structure of the double-arrow negative Poisson's ratio structure to further improve the crash performance of the energy-absorbing box and meet the crashworthiness of the car.

In this paper, a new double-arrow type negative Poisson's ratio energy absorption box is established on the basis of the front energy absorption box of a passenger car. The model of the original energy-absorbing box is shown in Fig. 5.1. The energy-absorbing box is composed of two parts welded together, and the surface of the parts is equipped with an induced groove to deform the structure in the designed way so as to increase the energy absorption capacity. The length of the original box is 202 mm, the width is 116 mm, the height is 62 mm, the material of the structure is SAPH440, the density of the material is 7850 kg/m^3, the modulus of elasticity is 210 GPa, and the Poisson's ratio is 0.3. The finite element model of the box is established in the general finite element pre-processing software Hyper mesh, and the model mainly consists of a quadrilateral mesh and a small number of the model mainly consists of a quadrilateral mesh and a small number of triangular meshes, the size of the mesh is about 6 mm $\times$ 6 mm, and the energy-absorbing box model contains 2987 cells and 3105 nodes in total.

Based on the dimensional parameters of the energy-absorbing box, a model of the double-arrow type negative Poisson's ratio structural energy-absorbing inner core is created. This model is depicted in Fig. 5.2c. The energy-absorbing inner core consists of an array of three-dimensional double-arrow negative Poisson's ratio structural cells in three directions. The double-arrow negative Poisson's ratio structural energy-absorbing inner core model contains four basic three-dimensional cells in the X direction, seven basic three-dimensional cells in the Y direction, and 24 in the Z direction. The three-dimensional double-arrow negative Poisson's ratio cells are shown in Fig. 5.2b, and the basic three-dimensional cells are combined by two two-dimensional arrow-shaped cells in two directions, and the two-dimensional double-arrow-shaped cells are shown in Fig. 5.2a.

The finite element model of the structure is established based on the three-dimensional numerical model of the negative Poisson's ratio energy-absorbing inner

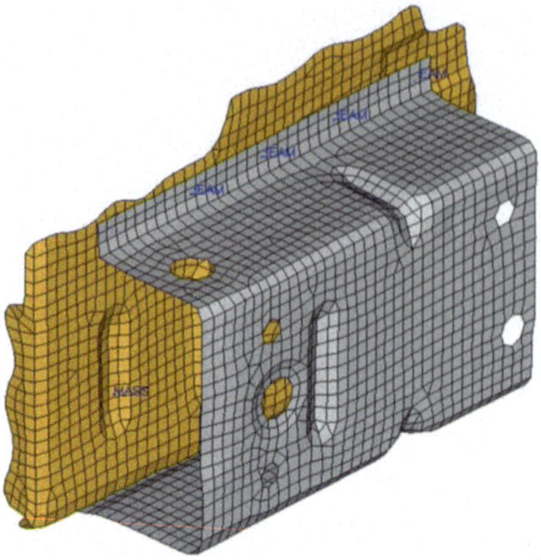

Fig. 5.1 Finite element model of traditional energy absorption box

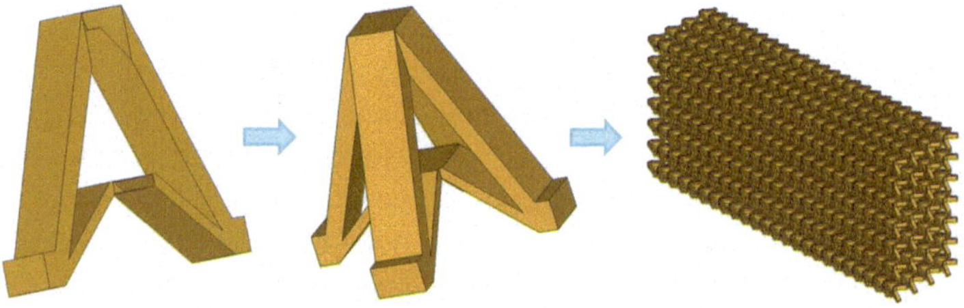

(a) 2D Negative Poisson's Ratio Structure
(b) 3D Negative Poisson's Ratio Structure
(c) Negative Poisson's Ratio Structure Energy Absorbing Inner Core

Fig. 5.2 Double arrow head type negative Poisson's ratio energy absorbing inner core

core of the double-arrow type, and the finite element is shown in Fig. 5.3a. Considering the contact between the three-dimensional negative Poisson's ratio structural units and simplifying the analysis, the connection between the negative Poisson's ratio structural units is merged by nodes. In order to improve the accuracy of the model analysis during the collision analysis, the energy-absorbing inner core model adopts quadrilateral mesh, and the energy-absorbing inner core finite element model contains 35,130 quadrilateral mesh cells and 36,541 nodes in total. By effectively connecting the double-arrow type negative Poisson's ratio energy-absorbing inner core and the initial energy-absorbing box, the finite element model of the new negative Poisson's ratio structure energy-absorbing box is established as shown in Fig. 5.3b.

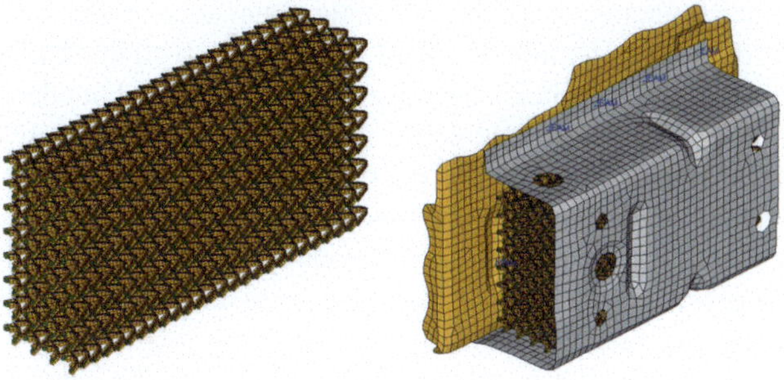

(a) Negative Poisson's Ratio Structural Energy Absorbing Inner Core
(b) Negative Poisson's Ratio Structural Energy Absorbing Box

Fig. 5.3 Double arrow head type negative Poisson's ratio structure energy absorption box

5.2.1.2 Study on the Performance of Double Arrowhead Negative Poisson Ratio Energy Absorbing Box

In order to study the performance of the double-arrow negative Poisson's ratio structure energy-absorbing box, the analysis of the energy-absorbing box was carried out by equating the structure with the low-speed crash conditions of the bumper according to the regulations. The energy-absorbing box structure was initially fixed at its bottom to constrain all six degrees of freedom. Subsequently, an impact was simulated by a rigid wall weighing 800 kg, striking the other end of the negative Poisson's ratio structure at an impact velocity of 16 km/h. The computational model was analyzed in the nonlinear finite element solver Ls-Dyna, and the kinetic and static friction factors of the structure were set to 0.2 and 0.3, respectively, while the friction factor between the structure and the impacting rigid wall was set to ensure high computational stability and efficiency, the contact between the energy-absorbing box and the rigid wall, as well as the contact between the box itself, are set as single-sided contact algorithms. The dynamic large-deformation nonlinear explicit finite element method was employed to solve the computation. To maintain accuracy, the time step used in the calculation was chosen to be smaller than the critical time step. This is because the central difference method utilized in the explicit finite element method is conditionally stable. For all cases in this study, the minimum time step employed was set as $7 \times 10^{(-4)}$ ms [21].

In order to compare the performance of the double-arrow negative Poisson's ratio structure energy-absorbing box, the analysis also includes the performance evaluation of the initial energy-absorbing box. Since the energy absorption of the box can be improved by filling the box with aluminum foam material in the previous designs of automotive manufacturers, the foam aluminum filled inner core energy absorption box is also established in this chapter as a comparison. According to the inner core size of the negative Poisson's ratio box structure, the original box is filled with aluminum foam with the same size as the negative Poisson's ratio structure, and the relevant material parameters of the aluminum foam are obtained by experiment, and the weight of the aluminum foam-filled inner core is 0.40 kg. The original box and the aluminum foam-filled box are analyzed under the same boundary conditions and load conditions, and Fig. 5.4 shows the original box and the negative Poisson's ratio structure. Poisson's ratio structural energy-absorbing box and aluminum foam-filled energy-absorbing box at different times during the collision. Comparing the deformation diagrams of the three different models at 12.49 ms, it can be seen that the different energy-absorbing box structures are the first to be crushed at the induced slot after the impact of the external force, because in the original energy-absorbing box structure, the induced slot is designed to produce controlled and stable plastic deformation and enhance energy absorption. Comparing the deformation diagrams of the three different models at t = 29.99 ms, it can be seen that the energy-absorbing box starts to collapse at the second induced slot, and the collapse pattern of the three different models remains the same, but the deformation of the energy-absorbing box is the largest at this time for the negative Poisson's ratio structure. Then until t = 60 ms, the energy-absorbing box is crushed at other places until the structure stops deforming, and when the collision is completed, the deformation patterns of the three still remain basically the same, and the final deformation of the original energy-absorbing box is the largest among the three models, and the foam aluminum-filled energy-absorbing box is the second. The collision process of the models is analyzed to obtain the parameters of total energy absorption, crush displacement and average collision force during the collision of the three different energy-absorbing boxes, and the comparison of different parameters during the collision of the three models is shown in Table 5.1.

Analysis of the above data shows that the masses of the double-arrow negative Poisson's ratio structure energy-absorbing box and the foam aluminum-filled energy-absorbing box are larger than the original energy-absorbing box due to the increase of the filled inner core, and the foam aluminum-filled energy-absorbing box has the largest mass. The energy-absorbing box is collapsed during the collision, and the collapse displacement of the double-arrow negative Poisson's ratio energy-absorbing box is the smallest, followed by the aluminum foam-filled energy-absorbing box, and the original energy-absorbing box is the largest. Since the energy absorbed by the three models is basically equal, the collapse displacement of the double-arrow negative Poisson's ratio structure energy-absorbing box is the smallest, which means that the double-arrow negative Poisson's ratio structure energy-absorbing box needs the smallest displacement to absorb the same energy when applying different three energy-absorbing boxes for collision, so the negative Poisson's ratio structure energy-absorbing box has a stronger energy absorption capacity per unit displacement length. In the context of vehicle structure design, the use of the double-arrow negative Poisson's ratio structure energy-absorbing box allows for the shortening of the length of the energy-absorbing box, which can provide more space for occupant compartment and further lightweight design.

Figure 5.5 shows the collision force–time curves of the original energy-absorbing box, the double-arrow negative Poisson's ratio structure energy-absorbing box and the aluminum foam-filled energy-absorbing box. It can be seen from the figure that

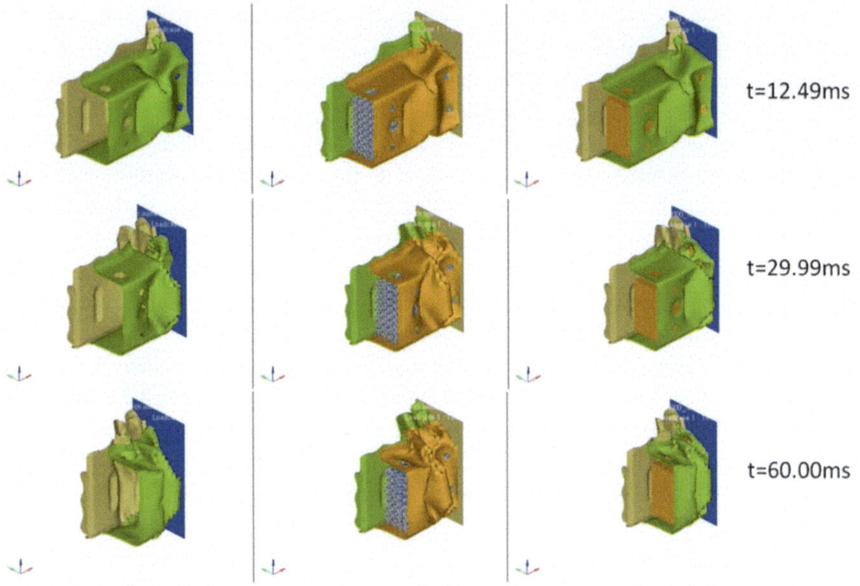

Fig. 5.4 Collapse deformation diagram of different energy absorption box models

Table 5.1 Comparison of collision process parameters of different energy absorption box models

	Original crash box	NPR crash box	Aluminum foam filled energy absorbing box
Total mass/kg	1.14	1.46	1.54
Total energy/kJ	7310	7330	7350
Displacement/mm	124.55	93.43	117.82
Average impact force/kN	58.69	78.45	62.38

the magnitude of the collision force in the initial stage of the collision is basically the same in the three models, which is mainly because in the initial stage of the collision, the energy-absorbing box sheet metal is mainly in contact, and the negative Poisson's ratio structure core and the foam aluminum core are not playing a role for the time being. At the first peak of the collision force, the peak collision force of the double-arrow negative Poisson's ratio structure energy-absorbing box is significantly larger than that of the aluminum foam-filled energy-absorbing box and the original energy-absorbing box, and the average collision force between the first peak and the second peak of the energy-absorbing box is larger than that of the other two energy-absorbing boxes. Before the end of the collision, the collision force of the double-arrow negative Poisson's ratio structure energy-absorbing box decreases sharply, which is because the collision energy is basically absorbed during the collision and the collision process is near the end. The average collision force of the three models is shown in Table 5.1, which shows that the average collision force of the double-arrow negative Poisson's ratio energy-absorbing box is the largest, the foam aluminum filled energy-absorbing box is the second largest, and the average collision force of the original energy-absorbing box is the smallest. In the design of the energy-absorbing box of the car, the first peak and the average collision force are generally expected to increase within a reasonable range, so that the structure has the ability to absorb more energy when the box is crushed. The first peak collision force and the average collision force of the aluminum foam-filled energy-absorbing box are larger than the original energy-absorbing box, so the aluminum foam-filled energy-absorbing box also has better energy absorption capacity than the original energy-absorbing box.

Figure 5.6 shows the variation curves of internal energy absorbed by the three models with time. Analyzing the curve data in the figure, it can be seen that the original energy-absorbing box absorbs 7310 kJ of internal energy during the collision, the double-arrow negative Poisson's ratio structure absorbs 7330 kJ of internal energy, and the foam aluminum-filled energy-absorbing box absorbs 7350 kJ of internal energy, and the internal energy absorbed by the three models is basically the same. Comparing the internal energy absorption curves of the three different energy-absorbing boxes, we can see that the trend of the curves can be roughly divided into three stages. The first stage is the time from the first contact of the box with the rigid wall to 10 ms, and the trend of the energy curves of the three models in this stage is basically the same. The second stage is from 10 to 40 ms, in this stage, as the deformation increases, the energy absorbed by the double-arrow negative Poisson's ratio structure energy-absorbing box at the same time is much larger than the energy absorbed by the original energy-absorbing box and the aluminum foam-filled energy-absorbing box, and the gap between the curves is the gap of energy

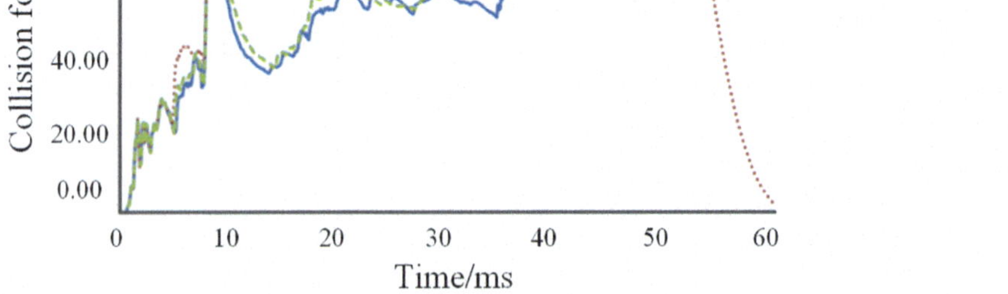

Fig. 5.5 Comparison of impact forces of different energy absorbing boxes

absorption between different models phase. The third stage is 40 ms until the end of the collision, in which the original energy-absorbing box and the aluminum foam-filled energy-absorbing box continue to absorb energy, while the negative Poisson's ratio structure energy-absorbing box basically no longer absorbs energy, mainly because the double-arrow negative Poisson's ratio structure energy-absorbing box has been completed in advance due to its strong energy absorption ability. During the whole collision process, the internal energy absorbed by the aluminum foam-filled energy-absorbing box and the original energy-absorbing box has basically the same trend with time, and the energy absorbed by the aluminum foam-filled energy-absorbing box is larger than that absorbed by the original energy-absorbing box within the time 34–39 ms, which means that the energy absorption capacity of the aluminum foam-filled energy-absorbing box is larger than that of the original energy-absorbing box during this time. From the above analysis, it can be seen that the energy absorption capacity of the negative Poisson's ratio structure is greater than that of the original energy-absorbing box and the aluminum foam-filled energy-absorbing box, while the energy absorption capacity of the aluminum foam-filled energy-absorbing box is also greater than that of the original energy-absorbing box.

In order to further compare and analyze the energy absorption characteristics of different energy-absorbing box models, the above three models were crushed at a constant speed of 5 m/s, and the crush displacement was 160 mm. Figure 5.7 shows the collision force–time curves of three different energy-absorbing box models. From the figure, it can be seen that the collision force of the aluminum foam-filled energy-absorbing box is the largest among the three models in the initial stage of collision, so it absorbs the most energy. However, as the collision proceeds, the collision force of the double-arrow negative Poisson's ratio structure rises rapidly, and at the first peak the double-arrow negative Poisson's ratio structure energy-absorbing box has the largest peak collision force, followed by the aluminum foam-filled energy-absorbing

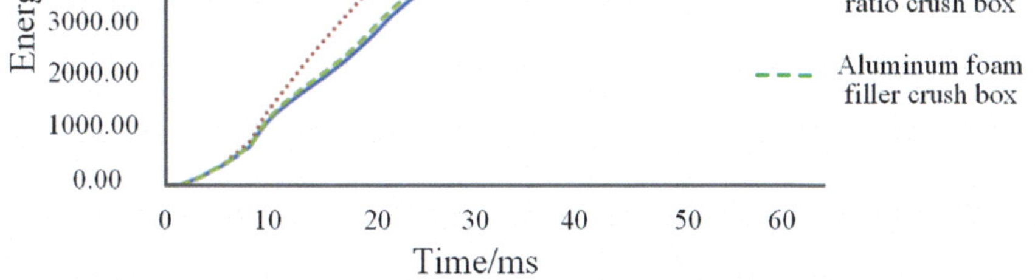

Fig. 5.6 Energy absorption change curve of different energy absorption boxes

box, and the original energy-absorbing box has the smallest. After the first peak, the collision force between the aluminum foam-filled box and the original energy-absorbing box decreases rapidly, and the collision force between the two boxes is basically the same, while the collision force of the double-arrow negative Poisson's ratio energy-absorbing box remains at a higher level, and its collision force is much larger than that of the other two energy-absorbing boxes. Because the model performs uniform collision analysis, it is known from the definition of energy absorption that the double-arrow negative Poisson's ratio structural energy-absorbing box will absorb more energy in the collision. The second peak of the collision force of the double-arrow negative Poisson's ratio structure is much larger than the second peak of the other two models in Fig. 5.7. This is because the double-arrow negative Poisson's ratio structure will undergo densification strain during the collision, which will further increase the energy absorption capacity of the structure.

The comparison of the internal energy absorption curves of the three models during the uniform velocity collision is shown in Fig. 5.8. Comparing the internal energy absorption curves, it can be seen that the energy absorbed by the aluminum foam-filled energy-absorbing box is more in the initial collision phase among the three models, but as the collision proceeds, the energy absorbed by the double-arrow negative Poisson's ratio structure energy-absorbing box is gradually larger than that absorbed by the other two energy-absorbing boxes, and the difference between the energy absorbed by different models increases with the increase of time. During the collision, the energy absorbed by the aluminum foam-filled box is also larger than that absorbed by the original box, but the difference between them is much smaller than that between the original box and the double-arrow negative Poisson's ratio box. The above analysis shows that the double-arrow type negative Poisson's ratio absorbing box has the largest energy absorption capacity, followed by the aluminum foam filled absorbing box, and the original absorbing box is the smallest.

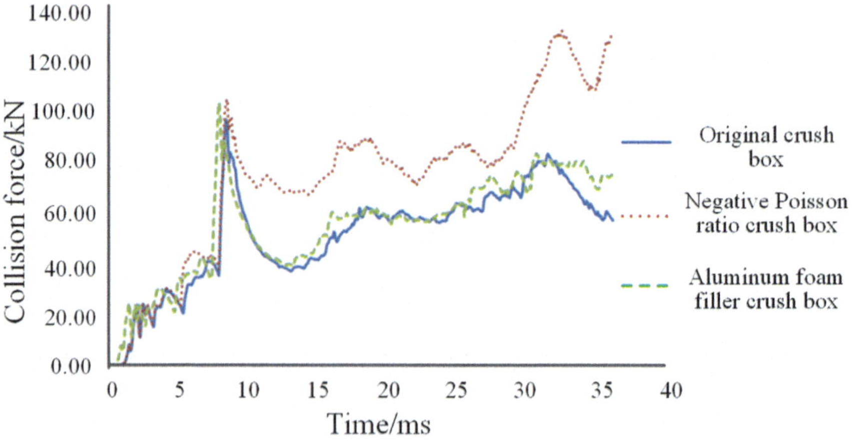

Fig. 5.7 Comparison diagram of collision forces of different energy absorbing boxes in uniform velocity collision

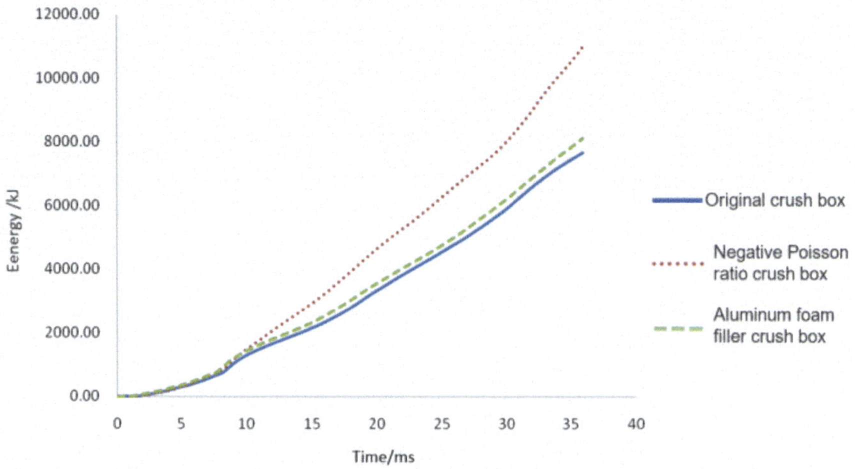

Fig. 5.8 Internal energy absorption curve of different energy absorbing boxes during uniform velocity collision

Considering the lightweight design principle in the design process, the specific energy absorption of the three models is compared, and Fig. 5.9 shows the comparison of the specific energy absorption curves of the three models in the collision process. In the initial stage of the collision, although the energy absorbed by the aluminum foam-filled energy-absorbing box is more, its specific energy absorption is still smaller due to its larger mass. As the collision proceeds, the energy absorbed by the double-arrow negative Poisson's ratio box is gradually larger than that of the other two boxes, and although its mass is larger than that of the original box, its specific absorption energy is gradually larger than that of the original box as the collision proceeds, and the gap between the models becomes larger as time increases. Compared with the aluminum foam-filled box, the double-arrow negative Poisson's ratio energy-absorbing box has a relatively lighter mass and absorbs more energy than the aluminum foam-filled box, so there is a larger difference between the two energy-absorbing boxes during the collision.

Figure 5.10 shows the deformation diagram of the energy-absorbing inner core of the double-arrow type negative Poisson's ratio structure. From the figure, it can be observed the negative Poisson's ratio structure also shrinks in the other two directions after the core is crushed during the collision. This behavior confirms the presence of a negative Poisson's ratio effect in the structure. The double-arrow negative Poisson's ratio structure core is stretched in the other two directions because the core is connected to the original box during the energy-absorbing box modeling process, and the corresponding double-arrow negative Poisson's ratio structure unit is stretched when the original box induced deformation out of the slot. The deformation pattern of the double-arrow negative Poisson's ratio structural inner core is consistent with the previous analysis, thus further demonstrating that the negative

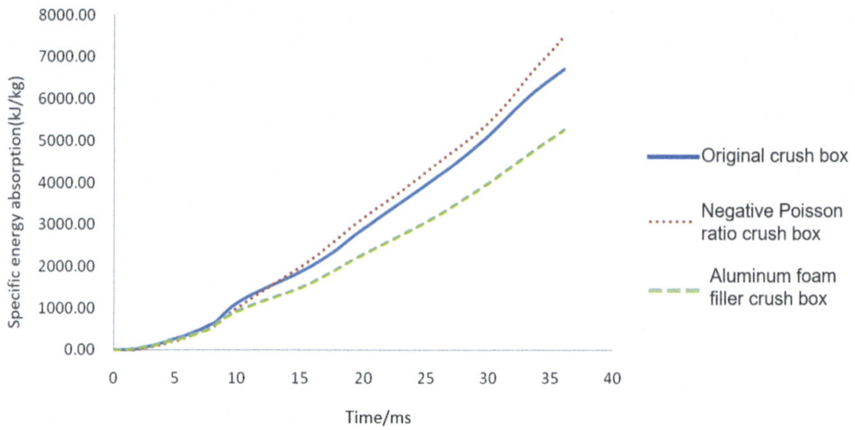

Fig. 5.9 Time dependent curve of specific energy absorption of energy absorption box during constant velocity collision

Poisson's ratio structural inner core can significantly improve the energy absorption performance of the energy-absorbing box.

In order to verify the effect of the change of energy-absorbing box performance on the crashworthiness of the whole vehicle [18] and to make a step-by-step comparison of the performance of the negative Poisson's ratio structural energy-absorbing box, three different energy-absorbing boxes are placed in the whole vehicle model of a

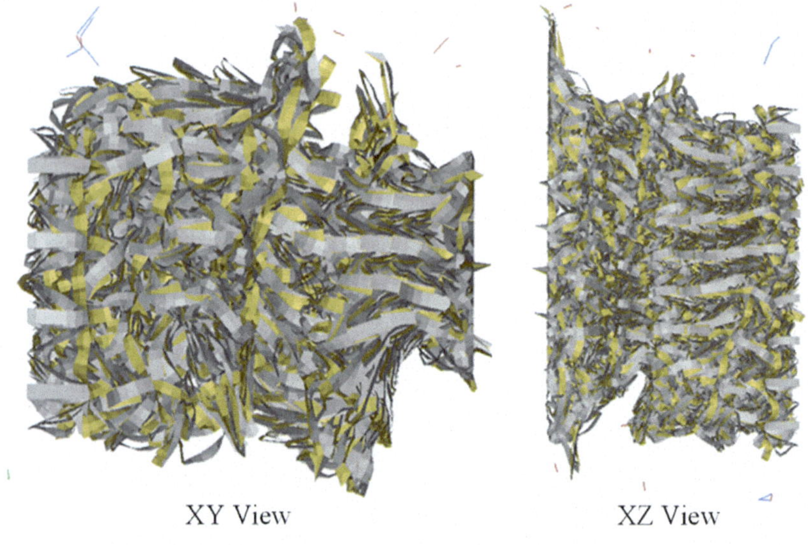

Fig. 5.10 Inner core deformation diagram of double arrow head type negative Poisson's ratio energy absorption box

Fig. 5.11 Whole vehicle collision model based on traditional energy absorption box

passenger car, and the whole vehicle crash models based on the double-arrow type negative Poisson's ratio structural energy-absorbing box, the conventional energy-absorbing box and the foam aluminum-filled energy-absorbing box are established in this paper. The crash model based on the conventional energy-absorbing box is shown in Fig. 5.11. The conventional energy-absorbing box model contains 1.65 million units and 1.66 million nodes. The analysis is carried out for a 100% overlap frontal collision with a crash speed of 50 km/h.

Figure 5.12 shows the comparison curves of the B-pillar acceleration curves during the collision of the three models, the peak B-pillar acceleration is important for the crashworthiness of the car and is an important index for its evaluation, generally the lower the peak, the better the occupants can be protected, and the peak of the B-pillar acceleration curve should be reduced as much as possible in the crash design [22]. Analyzing the curve data in the figure, it can be seen that among the three curves, the peak B-pillar acceleration of the whole-vehicle crash models based on double-arrow negative Poisson's ratio structural energy-absorbing box, conventional energy-absorbing box and foam aluminum-filled energy-absorbing box are 41.6, 44.4 and 45.4 g, respectively, and the peak B-pillar acceleration of the whole-vehicle crash model with double-arrow negative Poisson's ratio structural energy-absorbing box is the smallest, which is 6.31% lower than that of the whole-vehicle crash model with conventional energy-absorbing box. The peak acceleration of B-pillar of the whole-vehicle crash model with aluminum foam-filled energy-absorbing box is slightly higher than that of the whole-vehicle crash model with conventional energy-absorbing box. Therefore, it can be seen that by applying the double-arrow negative Poisson's ratio material energy-absorbing box instead of the conventional energy-absorbing box, the crashworthiness of the whole vehicle is significantly improved.

The above analysis shows that compared with the original energy-absorbing box and the aluminum foam-filled energy-absorbing box, the double-arrow negative Poisson's ratio structure energy-absorbing box can not only significantly improve the energy absorption capacity of the box, but also increase the energy absorption capacity per unit mass of the box structure. Therefore, by applying the double-arrow negative Poisson's ratio structure energy-absorbing box in automobile crash design, it not only can reduce the damage to the occupant compartment in the crash, but also can make the structure further lightweight design, which is important for improving the automobile crash performance and reducing the automobile production cost.

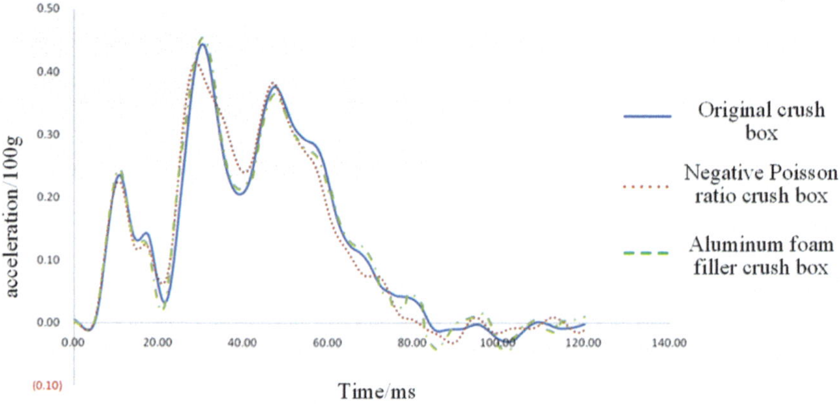

Fig. 5.12 Comparison of B-pillar acceleration curves of complete vehicle collision models based on different energy absorption boxes

5.2.1.3 Effect of Microstructure Parameters on the Performance of Double Arrow Shaped Negative Poisson's Ratio Energy Absorbing Box

From the above analysis, it is clear that, in addition to the basis material and the composition of the arrow-shaped cells, the microscopic topology of the negative Poisson's ratio structure has an important influence on the macroscopic performance of the structure, and the performance of the negative Poisson's ratio structure composed of arrow-shaped cells will have a large variation when the arrow-shaped cells have different cell angle, cell wall thickness and cell wall length. In this subsection, the effects of different cell angle, cell wall thickness and cell wall length on the performance of the double-arrow negative Poisson's ratio structural energy-absorbing box are investigated when the substrate material and the composition of the arrow-shaped cells of the double-arrow negative Poisson's ratio structural energy-absorbing box are determined, so as to provide design guidance for the further application of the negative Poisson's ratio structural energy-absorbing box in automotive design and manufacturing.

Effect of Cell Angle on the Performance of Energy Absorbing Box with Negative Poisson's Ratio Structure

The angle of the arrow-shaped meta cell structure can be categorized into two types: the angle between the long cell wall and the geometric symmetry axis, and the angle between the short cell wall and the geometric symmetry axis. The analysis of the composition of the arrow-shaped meta cell structure shows that when the lengths of the long and short cell walls are kept constant, the angle θ between the long cell wall and the angle φ between the short cell wall affect each other, and when the angle θ

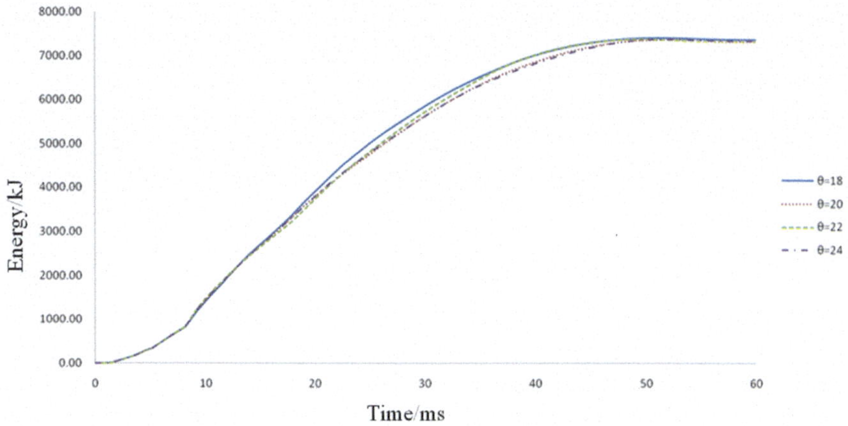

Fig. 5.13 Energy absorption curve of energy absorption box with different wall angles

increases, the angle φ also increases due to the influence of the cell wall length and vice versa. By varying the angle θ, four negative Poisson's ratio energy-absorbing boxes with different angles of 18, 20, 22 and 24° are established. Except for the different cell wall angles, the substrate material, organization and other parameters of the different energy-absorbing box models remain the same as the original energy-absorbing box structure, and the same boundary conditions and loads are applied to the four models during the analysis.

Figure 5.13 shows the energy absorption curves of the double-arrow type negative Poisson's ratio structure energy-absorbing box models with different long cell wall angles, and the analysis of the curves shows that the energy absorbed by the negative Poisson's ratio structure energy-absorbing box with different angles is basically the same in the initial and final stages of the collision, which is because the middle stage of the collision is the main stage of energy absorption during the collision, so the influence of the change of the angle on the performance is mainly reflected in this stage. In this phase, the energy absorbed by the negative Poisson's ratio box is the most when the cell wall angle $\theta = 18°$. The energy absorbed by the negative Poisson's ratio box with the cell wall angle $\theta = 22°$ is smaller than that absorbed by the boxes with the cell wall angle $\theta = 20°$ and $\theta = 24°$ in the beginning phase, but when the deformation increases, the energy absorbed is gradually larger than that absorbed by the other two models.

The comparison of the parameters of the negative Poisson's ratio structured energy-absorbing box with different cell wall angles in the collision process is shown in Table 5.2. Analyzing the data in the table, it can be seen that the weight of the negative Poisson's ratio structural energy-absorbing box gradually increases when the angle of the cell wall decreases, which is due to the fact that the number of arrow-shaped cells gradually increases in order to keep the length of the inner core consistent in the collision direction. The crush displacement of the negative Poisson's ratio structural energy-absorbing box is the smallest when $\theta = 22°$ and the largest

Table 5.2 Comparison of parameters during collision of energy absorbing boxes with different sibling wall angles

Cell wall angle	$\theta = 18°$	$\theta = 20°$	$\theta = 22°$	$\theta = 24°$
Mass/kg	1.51	1.49	1.48	1.46
Crushing displacement/mm	88.28	91.86	87.54	93.43
Total energy absorption/kJ	7401	7364	7383	7330
Average impact/kN	83.84	80.17	84.34	78.45
Specific energy absorption/(kJ/kg)	4901	4942	4989	5021

when $\theta = 24°$. The total energy absorbed during the collision remains basically the same for the four different models.

Table 5.2 also shows the average collision force and specific absorption energy of different models to further analyze the performance of energy-absorbing boxes with different cell wall angles. The average collision force of the box does not have a direct linear relationship with the cell wall angle, and the average collision force of the negative Poisson's ratio box is the largest when $\theta = 22°$, which means that the negative Poisson's ratio box has the largest energy absorption capacity among the four different models and takes the shortest time to absorb the same amount of energy. With the increase of the cell wall angle, the specific energy absorption of the negative Poisson's ratio box gradually becomes larger, and when $\theta = 24°$, the specific energy absorption of the box is the largest, so the structure absorbs the largest energy per unit mass in the four models.

Effect of Cell Wall Thickness on the Performance of Energy Absorbing Boxes with Negative Poisson's Ratio Structures

The thickness of the cell wall of the arrow-shaped meta cell in the negative Poisson's ratio structure has a significant effect on the performance and weight of the structure, so the design of the cell wall thickness should be considered in the design process. The thickness of the cell wall in the arrow-shaped cell mainly includes the long cell wall thickness and the short cell wall thickness, and the effects of the long and short cell wall thickness on the performance of the double-arrow negative Poisson's ratio structure are studied in this subsection.

In this subsection, four different models with long cell wall thicknesses of 0.7, 0.8, 0.9 and 1.0 mm are built with the initial cell wall thickness as the reference, keeping the other parameters in the double-arrow type negative Poisson's ratio structural energy absorption box model consistent. The same constraints and loads are applied to the models with different cell wall thicknesses, and the energy absorption curves of the different models are shown in Fig. 5.14. It can be seen from the figure that the four different energy-absorbing box models have the same trend during the collision, especially in the initial and final stages of the collision, and the energy absorption curves of the different models remain basically the same. From 10 to 50 ms, the

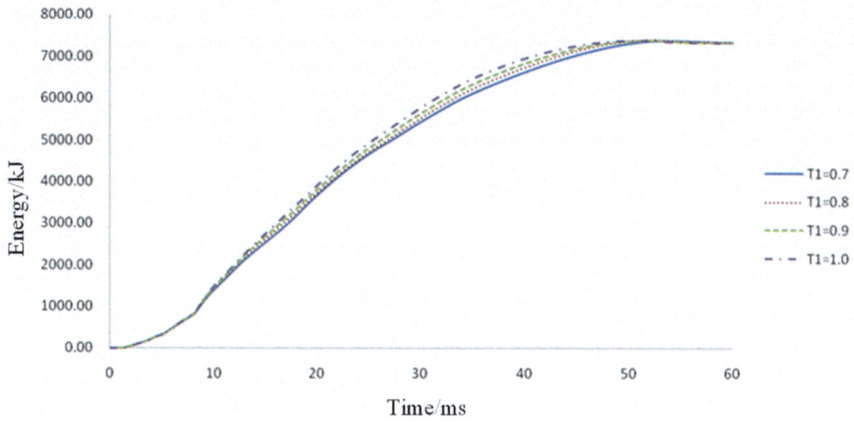

Fig. 5.14 Energy absorption curves of energy absorption boxes with different long cell wall thicknesses

Table 5.3 Comparison of parameters during collision of energy absorbing boxes with different long cell wall thicknesses

	$T_l = 0.7$ mm	$T_l = 0.8$ mm	$T_l = 0.9$ mm	$T_l = 1.0$ mm
Mass/kg	1.40	1.43	1.46	1.49
Crushing displacement/mm	97.69	94.35	93.43	90.71
Total energy absorption/kJ	7387	7393	7330	7380
Average impact /kN	75.62	78.36	78.45	81.36
Specific energy absorption/(kJ/kg)	5276	5170	5021	4953

double-arrow negative Poisson's ratio structure with long cell wall thickness $T_l = 1.0$ mm absorbs the most energy, and the double-arrow negative Poisson's ratio structure with long cell wall thickness $T_l = 0.7$ mm absorbs the least energy.

The energy absorbed by the four models with different long cell wall thicknesses and double-arrow Poisson's ratio energy-absorbing box during the collision is shown in Table 5.3. Comparing the data in the table, it can be seen that the total energy absorbed by the four models is basically the same, and the energy absorbed by the negative Poisson's ratio energy-absorbing box with $T_l = 0.8$ mm is the most. The collapse displacement of the negative Poisson's ratio energy-absorbing box with the long cell wall thickness $T_l = 1.0$ mm is the smallest among the four models after the collision, while the required collapse displacement of the model with $T_l = 0.7$ mm is the largest, and it can be seen that the collapse displacement decreases gradually with the increase of the long cell wall thickness.

For further comparative analysis of the performance of the energy-absorbing box, the average collision force and specific energy absorption of the four models are listed in Table 5.3. Analyzing the data in the table, it can be seen that the average collision force of the double-arrow negative Poisson's ratio structure energy-absorbing box

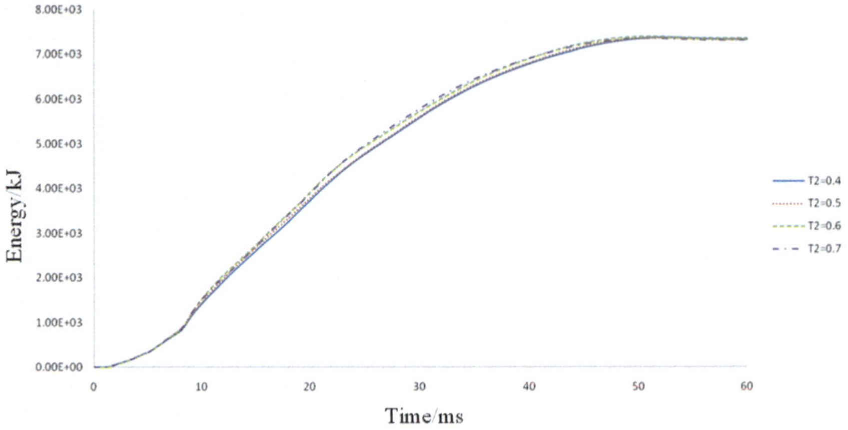

Fig. 5.15 Energy absorption curves of energy absorption boxes with different short cell wall thicknesses

increases with the increase of the long cell wall thickness, so it can be seen that the energy absorption capacity of the double-arrow negative Poisson's ratio structure energy-absorbing box increases with the increase of the long cell wall thickness, but when the thickness of the long cell wall increases, the specific energy absorption of the box gradually decreases, and the energy absorbed per unit mass gradually decreases with the increase of the long cell wall thickness. The energy absorbed per unit mass decreases as the thickness of the long cell wall increases. Therefore, when designing the long cell wall thickness, the energy absorption capacity and specific absorption energy of the double-arrow negative Poisson's ratio structure absorbing box should be considered comprehensively.

In order to study the effect of short cell wall thickness on the energy absorption of the negative Poisson's ratio structural energy-absorbing box and to keep all other parameters in the model consistent, four different energy-absorbing box models with thicknesses of 0.4, 0.5, 0.6 and 0.7 mm were built with the initial thickness of the short cell wall as the reference. The same constraints and loads are applied to all models as in the previous analysis. Figure 5.15 shows the energy absorption curves of the energy-absorbing box models with different short cell wall thicknesses and negative Poisson's ratio structures. From the figure, it can be seen that the energy absorption curves of different short cell wall thickness models have the same trend in different stages of the collision process, especially in the initial and final stages of the collision. However, the absorbed energy of different models shows different in 10–50 ms time, and the negative Poisson's ratio structure model absorbs the most energy at $T_2 = 0.6$ mm and $T_2 = 0.7$ mm.

To further analyze the performance of the negative Poisson's ratio structural energy-absorbing box models with different short cell wall thicknesses, Table 5.4 shows the comparison of the parameters of the four different energy-absorbing box models. Comparing the total absorbed energy of the four models, it can be seen that

Table 5.4 Comparison of parameters during collision between different short cell wall thickness models

	$T_2 = 0.4$ mm	$T_2 = 0.5$ mm	$T_2 = 0.6$ mm	$T_2 = 0.7$ mm
Mass/kg	1.45	1.46	1.47	1.49
Crushing displacement/mm	94.17	93.43	89.5	87.38
Total energy absorption/kJ	7380	7330	7404	7354
Average impact/kN	78.37	78.45	82.73	84.16
Specific energy absorption/(kJ/kg)	5090	5021	5037	4936

the energy-absorbing box absorbs the most energy when $T_2 = 0.6$ mm, and the least energy is absorbed when $T_2 = 0.5$ mm. When the thickness of the short cell wall increases, the required crush displacement of the negative Poisson's ratio structure energy-absorbing box gradually decreases, so the length of the crush process can be reduced when the negative Poisson's ratio structure energy-absorbing box has a larger thickness of the short cell wall during the collision.

In order to compare the energy absorption characteristics and the lightweighting characteristics of the energy-absorbing box, the average collision force and specific energy absorption data of different models are also compared in Table 5.4. Analysis of the data shows that the average collision force of the negative Poisson's ratio structured energy-absorbing box increases with the increase of the thickness of the short cell wall, so when the structure has a larger thickness of the short cell wall, its energy absorption capacity is relatively stronger. The specific energy absorption of the negative Poisson's ratio structure does not have a direct linear relationship with the thickness of the cell wall, and the specific energy absorption of the box is the largest when $T_2 = 0.5$ mm, so the energy absorbed per unit mass is the largest.

The above analysis shows that the variation of the thickness of the long and short cell walls has a significant effect on the mass, energy absorption and collision force of the negative Poisson's ratio structure energy-absorbing box. The design of the structure should be based on the performance requirements and cost of the structure, so as to design a reasonable thickness of the long and short cell walls, so that the structure has the energy absorption capacity, crush displacement and mass according to the design requirements.

Effect of Cell Wall Length on the Performance of Energy Absorbing Boxes with Negative Poisson's Ratio Structures

The microscopic shape of the arrow-shaped cell in the negative Poisson's ratio structure is determined by the parameters l_1, θ, φ, etc. When the angle between the cell walls of the arrow-shaped cell remains constant, the change of the long cell wall length l_1 will cause the corresponding change of the short cell wall length l_2, so in this paper, when studying the effect of the cell wall length on the performance of the negative Poisson's ratio structure energy-absorbing box, the parameters θ, φ,

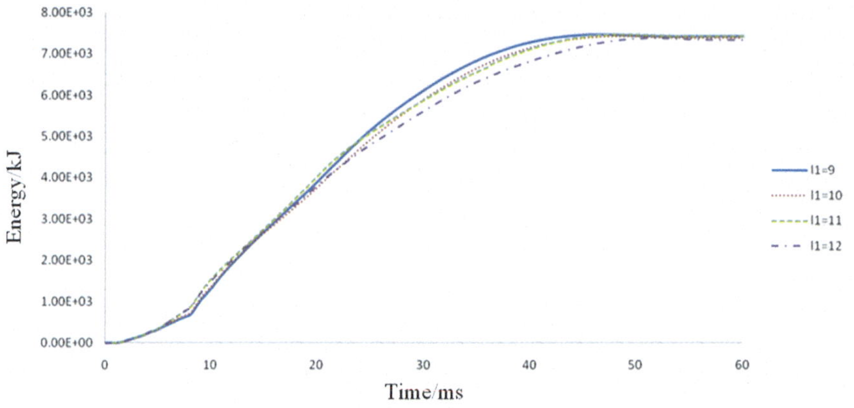

Fig. 5.16 Energy absorption curve of energy absorption box with different wall lengths

etc. are kept constant, and only the effect of the change of the long cell wall length on the performance of the negative Poisson's ratio structure energy-absorbing box is studied. In this subsection, the initial length of the long cell wall is used as the reference, and four different models with cell wall lengths of 9, 10, 11 and 12 mm are established. In order to keep the dimensions of the negative Poisson's ratio structural energy-absorbing core in the collision direction constant during the study, the number of negative Poisson's ratio structural cells in the collision direction changes accordingly when the cell wall length changes.

Figure 5.16 shows the energy absorption curves of energy-absorbing boxes with different cell wall lengths in the collision process, from which it can be seen that the energy absorption trends of different structures are different when the cell wall lengths are changed. In the initial stage of the collision, the energy absorption capacity of the box with a long cell wall length of 11 mm is larger than that of the box with a long cell wall length of 9 mm, but as the deformation increases, the energy absorbed by the box with a long cell wall length of 9 mm is the largest in the four models until the structure stops deforming. Analyzing the data, it can be seen that the same trend exists for other negative Poisson's ratio absorbing boxes with different cell wall lengths, i.e., the short cell wall absorbing box absorbs relatively less energy at the initial stage of the collision, but as the deformation of the absorbing box increases, the energy absorbed is gradually greater than that absorbed by the longer cell wall absorbing box.

The masses of the four different models are compared in Table 5.5. The mass of the energy-absorbing box is the largest for a long cell wall length of 9 mm, which is because when the cell wall length becomes shorter, the length of the arrow-shaped cell structure becomes smaller in the collision direction, thus the number of negative Poisson's ratio cells becomes larger and the mass of the structure increases. When $l_1 = 9$ mm, the energy-absorbing box requires the smallest crush displacement among the four models, and the crush displacement of the energy-absorbing

Table 5.5 Comparison of parameters during collision of energy absorbing boxes with different wall lengths

	$l_1 = 9$ mm	$l_1 = 10$ mm	$l_1 = 11$ mm	$l_1 = 12$ mm
Mass/kg	1.56	1.49	1.48	1.46
Crushing displacement/mm	79.51	83.39	85.06	93.43
Total energy absorption/kJ	7468	7428	7475	7330
Average impact/kN	93.93	89.08	87.88	78.45
Specific energy absorption/(kJ/kg)	4787	4985	5051	5021

box increases with the increase of the cell wall length during the collision, which also means that the energy-absorbing box needs a longer space to absorb the collision energy when the cell wall length becomes larger. Among the four models with different negative Poisson's ratios, the energy absorbed by the box is the least when $l_1 = 12$ mm and the most when $l_1 = 11$ mm, and there is no direct linear relationship between the absorbed energy and the length of the cell wall.

In order to evaluate the energy absorption characteristics and the lightweight characteristics of the energy-absorbing box, Table 5.5 lists the average collision force and the specific energy absorption for different models. Analyzing the data in the table, it can be seen that the average collision force of the negative Poisson's ratio structure energy-absorbing box gradually decreases with the increase of the cell wall length, and the specific energy absorption increases with the increase of the cell wall length, so when the length of the inner core cell wall of the negative Poisson's ratio structure becomes smaller, the energy absorption capacity of the structure gradually increases but its capacity per unit mass gradually decreases. In the design process, it is necessary to consider the mass of the structure, the energy to be absorbed and the crush displacement to make the most suitable cell wall length for the energy absorbing box.

Effect of Cell Wall Width on the Performance of Energy Absorbing Boxes with Negative Poisson's Ratio Structures

In the negative Poisson's ratio structure, due to the small size of the arrow-shaped cell, the width of the long cell wall and the short cell wall are generally made equal in engineering practice in order to reduce the manufacturing process difficulty and scale up the application, so this paper only studies the case of the same width of both. Taking the initial width of the cell wall as the reference, four different models with widths of 2.3, 2.5, 2.7 and 2.9 mm were established to study the effect of different cell wall widths on the performance of the negative Poisson's ratio absorbing box by applying the same boundary conditions and loads to them.

Figure 5.17 shows the energy absorption curves of the energy-absorbing box with different cell wall widths. It can be seen from the figure that the energy absorption curves of different models have the same trend during the collision process. box

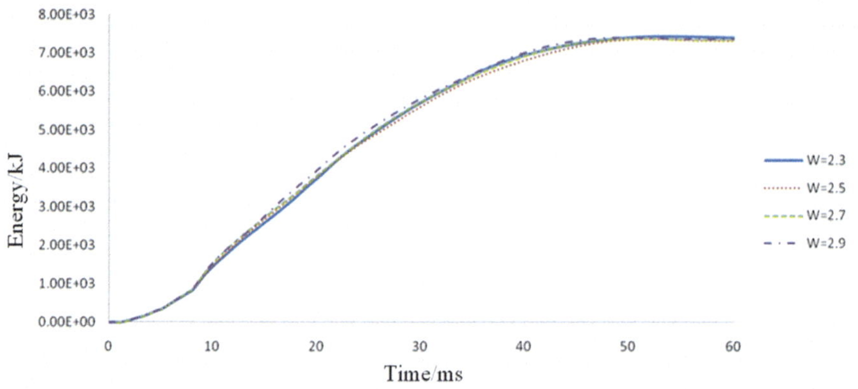

Fig. 5.17 Energy absorption curve of energy absorption boxes with different cell wall widths

Table 5.6 Comparison of parameters during collision of energy absorbing boxes with different wall widths

	$W = 2.3$ mm	$W = 2.5$ mm	$W = 2.7$ mm	$W = 2.9$ mm
Mass/kg	1.44	1.46	1.49	1.51
Crushing displacement/mm	95.16	93.43	88.74	88.03
Total energy absorption/kJ	7431	7330	7390	7409
Average impact/kN	78.09	78.45	83.29	84.16
Specific energy absorption/(kJ/kg)	5160	5021	4960	4907

takes the shortest time. Specifically, during the 10–50 ms energy absorption stage, the curve corresponding to the model with a width of W = 2.9 mm for the negative Poisson's ratio energy-absorbing box is positioned above the curves of the other three models. This indicates that the energy absorbed by the structure with a width of W = 2.9 mm is the highest, resulting in the shortest time required to absorb the same amount of energy during the collision process.

The comparison of the parameters during the collision of the four energy-absorbing box models with different widths is shown in Table 5.6. It can be seen that the negative Poisson's ratio energy-absorbing box with W = 2.3 mm absorbs the most energy during the collision and produces the largest crush displacement among the four models, while the energy-absorbing box with W = 2.5 mm absorbs the least energy. Comparing the crush displacements of different models, it can be seen that the crush displacements generated by the energy-absorbing box gradually decrease with the increase of the cell wall width.

The average collision force and specific energy absorption of each energy-absorbing box were calculated according to defined criteria and are presented in Table 5.6. Analyzing the data in the table reveals that as the width of the cell wall of the negative Poisson's ratio structure energy-absorbing box increases, the average collision force gradually rises during the collision. Additionally, the energy absorption capacity of the model increases with an increase in cell wall width. When comparing the specific energy absorption data of the models, it can be observed that the specific energy absorption of the negative Poisson's ratio structural energy-absorbing box gradually declines as the cell wall width increases. The specific energy absorption is the lowest when the width W = 2.9 mm. Therefore, it can be seen that different properties such as energy absorption and specific energy absorption of the negative Poisson's ratio structure absorbing box have different performance with the change of the cell wall width, and the width of the cell wall should be reasonably designed according to the requirements of the crush displacement, energy absorption and the quality of the structure, so that the absorbing box can achieve a better balance between different performance requirements.

5.2.1.4 Effect of Basis Materials on the Performance of Energy Absorption Boxes with Negative Poisson's Ratio Structures

The performance of the negative Poisson's ratio structured energy-absorbing box is not only influenced by the micro-topology of the arrow-shaped meta structure cell wall angle, width, and thickness but also relies heavily on the mechanical properties of the basis material used in the structure. The basis materials of the negative Poisson's ratio structure mainly include different types of steel, aluminum, plastic and composite. When the negative Poisson's ratio structure energy-absorbing box is impacted by external load, the long cell wall and short cell wall begin to produce bending deformation and thus show the linear elastic phenomenon, when the external load exceeds the load-bearing limit of the structure, the negative Poisson's ratio structure will occur according to the different basis materials such as brittle fracture and plastic collapse phenomenon. For elastoplastic materials, the deformation is first elastic stage, but when the stress exceeds the yield limit, plastic deformation of the material occurs, and the resulting deformation is not recoverable. Plastic materials are usually capable of large elastic deformation, but because there is no significant yield limit, the material will fracture directly at higher loads. Therefore, for a negative Poisson's ratio structure with the same topology and organization of arrow-shaped cells, different basis materials will exhibit different properties. The specific design process should be based on the selection of suitable basis materials for different applications and performance design requirements.

In this subsection, based on the same parameters of the negative Poisson's ratio structure energy-absorbing box, ordinary steel (BLC), high-strength steel (B280VK), aluminum and polymer material polypropylene (PP) plastic were selected as the basis materials to study the influence of different basis materials on the structural performance. The collisions were all fixed at one end of the energy-absorbing box, and then a rigid wall of 800 kg hit the other end of the box with an impact speed of 16 km/h.

Table 5.7 shows the comparison of the parameters of the four different models during the collision. From the above table, it can be seen that the negative Poisson's ratio structure energy-absorbing box with steel core (BLC and B280VK) has the largest weight and the PP core energy-absorbing box has the lightest mass. Comparing the total absorbed energy of the models, it can be seen that the energy absorbed by the absorbing box with ordinary steel (BLC) core during the collision is the most, the energy absorbed by the absorbing box with high strength steel (B280VK) core is slightly lower, and the total energy absorbed by the absorbing box with PP material core is the least. This also means that when applied to absorb the same amount of energy, the steel core box can be designed to have the smallest length, thus providing valuable design space for the vehicle. Figure 5.18 shows the variation curve of absorbed internal energy with time for four different models.

From the above figure, it can be seen that the energy absorbed by the four models in the initial stage of the collision is basically the same, and as the deformation increases, the energy absorbed by the ordinary steel and high-strength steel core energy-absorbing box continues to be the same and greater than the other two models, while the energy absorbed by the aluminum core energy-absorbing box is greater than that of the PP core energy-absorbing box. When all four models absorb the same amount of energy, the steel core box takes the least amount of time. Comparing the specific energy absorption of the different models in Table 5.7, it can be seen

Table 5.7 Comparison of parameters during collision between different basis material models

Base material	AL	PP	BLC	B280VK
Mass/kg	1.46	1.26	2.07	2.07
Crushing displacement/mm	93.43	110.16	72.68	72.71
Total energy absorption/kJ	7330	6996	7549	7519
Average impact/kN	78.45	63.51	103.87	103.41
Specific energy absorption/(kJ/kg)	5021	5552	3647	3632

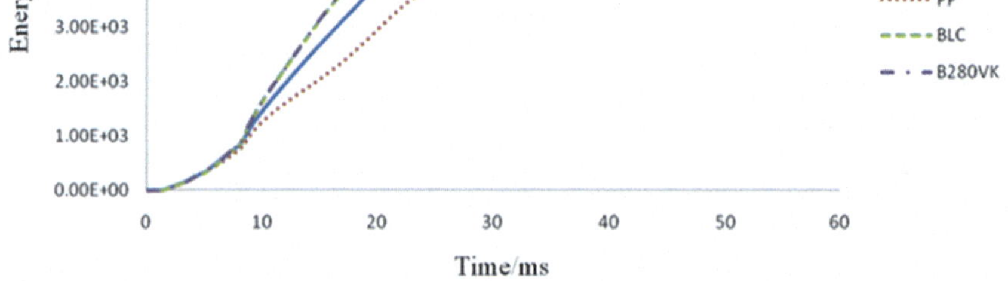

Fig. 5.18 Internal energy absorption curves of different basis material models

that although the energy absorbed by the PP-core box and the aluminum-core box is less, the specific energy absorption of both of them is much larger than that of the ordinary steel and the high-strength steel. The average collision force of the steel core energy-absorbing box is larger than the other two models, and the average collision force of the PP core energy-absorbing box is the smallest.

Based on the above analysis, it can be concluded that ordinary steel and high-strength steel, as the core basis materials of the energy-absorbing box, exhibit similar performance. During the collision process, the steel core energy-absorbing box experiences the highest average collision force and absorbs the most energy. The aluminum core energy-absorbing box follows, while the PP core energy-absorbing box exhibits the lowest performance. Furthermore, the steel core energy-absorbing box requires the smallest amount of crush displacement during the collision, which is beneficial for providing larger member space in automotive structures. However, the specific energy absorption of the aluminum and PP core energy-absorbing boxes is significantly higher than that of the steel core energy-absorbing box. Therefore, in the practical design of negative Poisson's ratio structure energy-absorbing boxes, the selection of basis materials needs to consider factors such as energy absorption, weight, and cost of the structure.

5.2.1.5 Effect of Impact Velocity on the Performance of Energy Absorbing Boxes with Negative Poisson's Ratio Structures

According to national regulations and C-NACP star crash requirements, cars need to be crashed in different forms at different speeds, and the crash speeds occurring during the driving process of the car have large differences, so the structural design and optimization process of the car needs to consider the design of the structure at different speeds, so that the structure can meet the requirements in a variety of different speed conditions [23]. For the negative Poisson's ratio structure, the equivalent platform stress of the material is closely related to the strain rate in addition to the substrate material, the shape and composition of the arrow-shaped cell, so when the negative Poisson's ratio structure energy-absorbing box is in different speed conditions, the strain rate will change in a wide range and the box will show different performance. This subsection will study the performance of the negative Poisson's ratio structure energy-absorbing box under different impact velocities to provide guidance for the application of the new energy-absorbing box in engineering practice. By applying the aluminum core energy-absorbing box, the collision analysis is performed at impact velocities of 12, 14, 16 and 18 km/h, respectively, while keeping the other constraints consistent.

The internal energy absorption curves of the model at different velocities are shown in Fig. 5.19. From the slope of the curves at different stages, it can be seen that the energy absorbed by the model at different velocities in the initial stage of the collision is less, which is mainly because the induced slot of the energy-absorbing box has not been crushed and the inner core of the negative Poisson's ratio structure has not been compressed, and no effective deformation has been carried out. After the initial stage, the model starts to absorb more energy, and the energy absorbed by the model gradually decreases towards the end of the collision until the end of the collision, when the energy-absorbing box no longer absorbs energy. Comparing the curves at different velocities, it can be seen that the time required from the start of collision to the completion of energy absorption in the negative Poisson's ratio structure energy-absorbing box is related to the initial collision velocity, the larger the velocity, the longer the time required to finish energy absorption, and vice versa. The comparison of absorbed energy, collapse displacement and average collision force of the model at different velocities is shown in Table 5.8. Analysis of the data in the table shows that as the initial velocity of the collision increases, the more energy is absorbed by the negative Poisson's ratio structural energy-absorbing box, and the required collision displacement is also larger.

For further comparison, Fig. 5.20 shows the comparison of the collision force of the energy-absorbing box with negative Poisson's ratio structure at different impact velocities. From the figure, it can be seen that the energy-absorbing box produces the highest impact velocity and takes the shortest time to reach the maximum collision force at V = 18 km/h. This is because the inertia effect is more obvious as the impact velocity increases. The second peak of the model gradually becomes larger and more

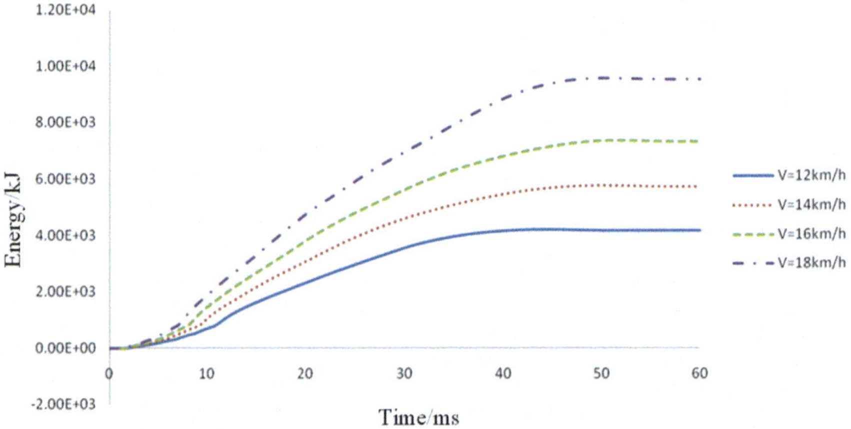

Fig. 5.19 Comparison of energy absorption of structures with negative Poisson's ratio at different speeds

Table 5.8 Parameter comparison of negative Poisson's ratio structural energy absorbing box under different impact speeds

Velocity	$V = 12$ km/h	$V = 14$ km/h	$V = 16$ km/h	$V = 18$ km/h
Crushing displacement/mm	48.71	73.27	93.43	109.81
Total energy absorption/kJ	4211	5773	7330	9574
Average impact/kN	86.45	78.79	78.45	87.19

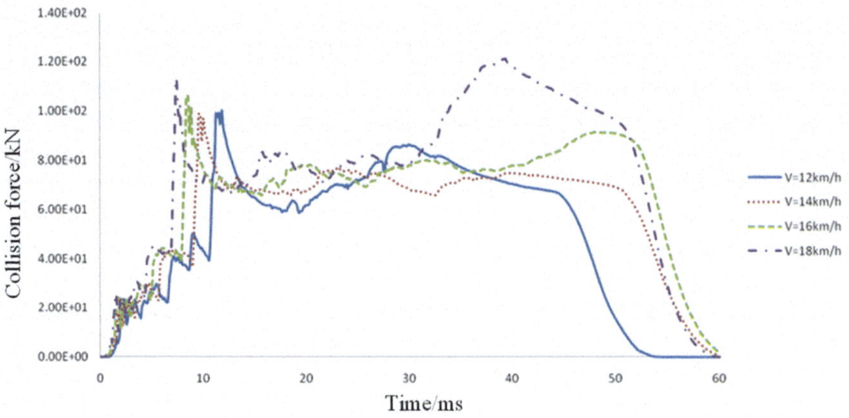

Fig. 5.20 Comparison of impact forces of negative Poisson's ratio structural energy absorbing boxes at different impact speeds

pronounced as the velocity increases, and this peak is related to the densification strain of the structure at larger impact velocities in addition to the increase in initial velocity. The comparison of the average collision force of the negative Poisson's ratio structural energy-absorbing box at different impact velocities is shown in Table 5.8. The model has the largest collision force at V = 18 km/h, the second largest at V = 12 km/h, and the smallest at V = 16 km/h. Thus, the average collision force of different models during the collision does not have a direct change pattern with the collision velocity.

The above analysis shows that with the increase of collision speed, although the average collision force of energy-absorbing box is not directly related to the collision speed, the inertia effect of collision becomes more obvious, the peak of collision force is bigger during the collision, the more energy absorbed by the negative Poisson's ratio structure of energy-absorbing box, the longer the required crush displacement is, and the structural design of energy-absorbing box should be guided according to the preliminary predetermined collision speed in the actual application of automobile.

5.2.1.6 Multi-objective Optimization of Energy Absorbing Box with Negative Poisson's Ratio Structure

The analysis of the effect of different parameters on the energy absorption box performance of the negative Poisson's ratio structure shows that the specific energy absorption of the negative Poisson's ratio structure gradually becomes larger with the increase of cell wall angle and cell wall length, and decreases with the increase of long cell wall thickness and cell wall width. The energy absorption capacity of the negative Poisson's ratio structure increases with the increase of long cell wall thickness, short cell wall thickness and cell wall width, and gradually decreases when the cell wall length increases. Therefore, the influence of the design parameters on the structural performance is beneficial for more targeted improvement of the structure, however, due to the large number of structural design parameters, the performance of the structure is related to the collision velocity and the substrate material, etc., and there is a nonlinear relationship between the variation of the parameters and some of the performance, especially when multiple targets need to be considered at the same time in the actual structural design process, the relationship between the performance of the structure and the parameters is more A more effective method is needed to design the energy-absorbing box for negative Poisson's ratio structures.

Based on the multi-objective optimization based on the parametric model in Chap. 4, this subsection proposes to first establish the parametric model of the negative Poisson's ratio structural energy-absorbing inner core, uses the node model generated by the parametric model and the traditional finite element model to generate the new negative Poisson's ratio structural energy-absorbing box model, then combine the DOE design method and the response surface method to establish the response surface model of the negative Poisson's ratio structural energy-absorbing box, and finally apply the NSGA-II multi-objective genetic algorithm is applied to find the multi-objective optimization solution of the negative Poisson's ratio structural energy-absorbing box. In this subsection, the negative Poisson's ratio structure is optimized by equivalence treatment in accordance with the national automobile low-speed crash regulations, and all the degrees of freedom of the nodes at one end of the energy-absorbing box are constrained during the analysis, and the other end of the energy-absorbing box is impacted with a rigid wall of 800 kg at an initial speed of 16 km/h.

The specific energy absorption (SEA) and the maximum collision force (PCF) are two important evaluation indexes for the performance of the energy-absorbing box. The design process should make the box have the maximum specific energy absorption and the smallest possible maximum collision force, so the specific energy absorption and the maximum collision force are used as the objective functions for the optimization of the negative Poisson's ratio structural energy-absorbing box in this section. Compared with the negative Poisson's ratio structure energy-absorbing inner core, the negative Poisson's ratio structure energy-absorbing box adds the traditional energy-absorbing box on top of it, and the traditional energy-absorbing box mainly consists of the energy-absorbing box plate and the cover plate, so two design variables, the thickness of the energy-absorbing box plate (T_3) and the thickness of the cover plate (T_4), are added to the negative Poisson's ratio structure energy-absorbing box for the optimization design. In the design process, the initial structure of the energy-absorbing inner core adopts the negative Poisson's ratio structure obtained from the multi-objective optimization in Sect. 4.5.3 of this chapter, so the design variables and the variation range in the optimization process of the negative Poisson's ratio structure are shown in Table 5.9.

In order to control the weight and cost of the negative Poisson's ratio structure energy-absorbing box, the weight of the negative Poisson's ratio structure energy-absorbing box in the optimization process cannot exceed 1.6 kg. Therefore, the mathematical model of the energy-absorbing box optimization process can be expressed

Table 5.9 Optimization design variables for structural energy absorbing boxes with negative Poisson's ratio

Design variables	Initialization value	Minimum value	Maximum value
l_l/mm	12.0	5	30
θ/rad	0.42	0	1.57
φ/rad	0.63	0	1.57
a/mm	2.5	1	15
T_1/mm	0.9	0.1	4
T_2/mm	0.5	0.1	4
T_3/mm	1.4	0.5	2
T_4/mm	1.4	0.5	2

by the above analysis as follows.

$$\begin{gathered} \text{Minimize } 1/\text{SEA, PCF} \\ \text{Subjected to:} \\ m \leq 1.6 \\ 0 < \theta < \varphi < 1.57 \\ 1 < a < 15, 5 < l_1 < 30 \\ 0.1 < T_1 < 4, 0.1 < T_2 < 4 \\ 0.5 < T_3 < 2, 0.5 < T_4 < 2 \end{gathered} \tag{5.4}$$

The parameter settings of the optimization algorithm during the optimization of the negative Poisson's ratio structured energy-absorbing box are consistent with those during the multi-objective optimization of the negative Poisson's ratio structured energy-absorbing inner core. After NSGA-II multi-objective optimization, 31 sets of negative Poisson's ratio structural energy-absorbing box Pareto non-inferior optimal solution sets are obtained, as shown in Fig. 5.29. It is known that the application of NSGA-II multi-objective genetic algorithm can provide multiple options for the design of negative Poisson's ratio structural energy-absorbing box, and the design process should be selected according to the specific design requirements (Fig. 5.21).

Table 5.9 lists five sets of typical Pareto optimization solutions for negative Poisson's ratio structural energy-absorbing box. Comparing different sets of optimization solutions in the table, it can be seen that the optimal solution of specific energy absorption and maximum collision force cannot be obtained simultaneously during the optimization of negative Poisson's ratio structural energy-absorbing box, and the actual design needs to be selected according to the specific design requirements. If you want to get a larger specific energy absorption, you should choose option 4; if you want to get a smaller maximum collision force for the negative Poisson's ratio structure energy-absorbing box, you should choose option 2 (Table 5.10).

According to the satisfaction function of Eq. (4.11) to find the satisfaction of each Pareto optimization solution, it can be seen that the solution 5 has the smallest

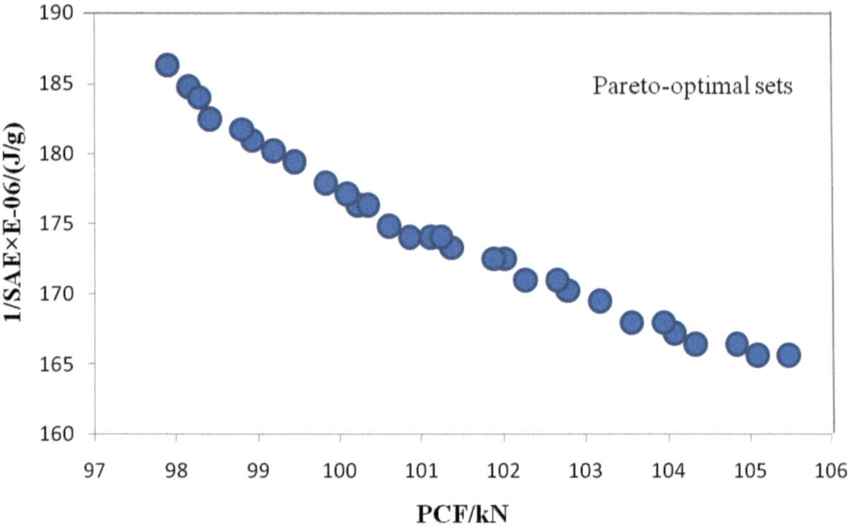

Fig. 5.21 Pareto non inferior optimal solution set for negative Poisson's ratio structural energy absorption boxes

Table 5.10 Comparison of pareto optimization solutions for negative Poisson's ratio structural energy absorption boxes

Design variables	Original scheme	Scheme1	Scheme2	Scheme3	Scheme4	Scheme1
l_l/mm	12.0	15.2	8.7	13.4	13.9	10.3
θ/rad	0.42	0.35	0.29	0.41	0.32	0.38
φ/rad	0.63	0.61	0.43	0.58	0.55	0.57
a/mm	2.5	2.2	1.9	1.4	3.1	2.8
T_1/mm	0.9	1.7	2.0	0.7	0.8	0.7
T_2/mm	0.5	0.6	0.8	0.8	0.5	0.6
T_3/mm	1.4	1.3	1.8	1.9	1.2	1.6
T_4/mm	1.4	1.8	1.1	1.3	1.7	1.1
SEA/J/g	5021	5859	5389	5658	5986	5802
PCF/kN	106.65	104.08	97.94	101.25	105.73	102.97

satisfaction value. Comparing the original solution and solution 5, it can be seen that the optimized negative Poisson's ratio structural energy-absorbing box has larger specific energy absorption and smaller maximum collision force, which satisfies the design constraints despite the increase in mass from 1.46 to 1.51 kg. The above analysis shows that the design parameters of the negative Poisson's ratio structure energy-absorbing box and the thickness of the conventional energy-absorbing box are optimized by using the negative Poisson's ratio structure energy-absorbing box optimization method proposed in this section, so that the performance of the negative Poisson's ratio structure energy-absorbing box can be further improved while satisfying the constraints. It should be noted that compared with the negative Poisson's ratio structure energy-absorbing core, the optimized solutions to meet the design requirements are different between the two due to the addition of different design variables and the influence of the position of the energy-absorbing core in the energy-absorbing box.

The energy-absorbing boxes play an important role in the crashworthiness design of automobiles, and the quality of their design significantly affects the crashworthiness performance of vehicles. In order to improve the energy absorption performance of the energy-absorbing box, this study proposes a new type of negative Poisson's ratio structure energy-absorbing box by effectively combining the negative Poisson's ratio structure energy-absorbing core with the traditional hollow energy-absorbing box. By establishing a collision model for the negative Poisson's ratio structure energy-absorbing box, comparisons are made with the traditional energy-absorbing box and foam-filled aluminum energy-absorbing box under three given conditions: initial velocity, constant collision velocity, and full-scale frontal collision. The analysis mainly focuses on energy absorption, collision force, specific energy absorption, crush displacement, and component mass, among other processes and parameters. The results show that the new type of negative Poisson's ratio structure energy-absorbing box exhibits better energy absorption characteristics and specific energy absorption during the collision process, substantially improving the crashworthiness performance of automobiles. Additionally, his paper also systematically investigates the effects of microstructural parameters such as cell wall angle, thickness, width and length, different crash speeds and different substrate materials on the performance of the negative Poisson's ratio structural energy-absorbing box, explores the effects of different parameters on the performance of the negative Poisson's ratio structural energy-absorbing box, provides guidance for the large-scale application of the negative Poisson's ratio structural energy-absorbing box in automotive design, and successfully realizes the negative Poisson's ratio structural energy-absorbing box for a certain vehicle model. The multi-objective optimization of the negative Poisson's ratio energy-absorbing box for a certain vehicle model was successfully achieved. The research results show that the new negative Poisson's ratio structural energy-absorbing box proposed in this chapter can effectively improve the crashworthiness of the vehicle and provide a new way to improve the performance of the energy-absorbing box, and the performance of the negative Poisson's ratio structural energy-absorbing box can be further optimized by reasonably optimizing the design microstructure parameters for different design requirements.

5.2.2 Application of Negative Poisson's Ratio Structure in Anti-collision Bar of Automobile Side Collision

5.2.2.1 Introduction

To provide better protection of passengers when the collision happens, promoting crashworthiness has been the focus of the vehicle design. Since side impact is one of the most important ways of vehicle collision, it is meaningful to improve the side impact crashworthiness of the vehicle. Unlike the vehicle frontal collision, there is little space for structures to deform and absorb the collision energy in the side impact. Furthermore, the side doors are usually the weakest part of the vehicle. These reasons make it much more challenging to improve the side impact crashworthiness. Side door impact beam is one of the most important parts which are used for resisting the collision and absorbing the collision energy in the side impact. The design to provide better protection of passengers when the collision happens, promoting crashworthiness has been the focus of the vehicle design. Since side impact is one of the most important ways of vehicle collision, it is meaningful to improve the side impact crashworthiness of the vehicle. Unlike the vehicle frontal collision, there is little space for structures to deform and absorb the collision energy in the side impact. Furthermore, the side doors are usually the weakest part of the vehicle. These reasons make it much more challenging to improve the side impact crashworthiness. Side door impact beam is one of the most important parts which are used for resisting the collision and absorbing the collision energy in the side impact. The design of the side door impact beam plays an important role in determining the side impact crashworthiness of the vehicle. Detailed studies which focus on finding an optimum cross-section shape of the side impact beam were conducted by Li to improve the energy absorption character. Lee designed a thin-walled side door beam using quenchable boron steel and geometry optimization is also performed via creating new reinforcing ribs to promote the performances. Yoon developed a new one-body door impact beam by applying the advanced high strength steel to improve the impact energy absorbability and proper cross section shape of the door impact beam was also obtained in the studies. These studies improved the performances of side door impact beam significantly. However, it is challenging to make further improvement on the performances of the side door impact beam due to the limitation of space and cost. By combining the traditional side door impact beam and NPR structure, a novel side door NPR impact beam is first proposed in this work to make further improvement of the side impact crashworthiness under the limitation of space. The performances of the side door NPR impact beam will be detailed studied by comparing to the traditional side door impact beam and aluminum foam filled impact beam in Sect. 5.2.

Based on the composition of the side door NPR impact beam, it can be concluded that the performances of the parts are affected by the shape of the traditional side door impact beam, parameters of the NPR structures and material types, etc. To maximize the potential of the side door NPR impact beam, it is meaningful to find the optimal design to provide better passenger protection. Optimization-based design methods have been applied successfully in the past years by a lot of researchers to improve performances and reduce weight. However, the structure optimization studies are usually based on the assumption that all the variables and design parameters are determined exactly and there is no uncertainty. In the practical engineering industry, there are certainly errors, which are caused by human or natural factors, in the design and manufacture. The optimal design obtained by the deterministic optimization will become unreliable or even wrong when the constraints are violated easily. Thus, the uncertainly of the variables should be considered in the optimization to make the optimization results more reliable. In this work, the multi-objective design optimization of the side door NPR impact beam with considering reliability is performed to improve the performances. The parametric model of the NPR structure is first established to enhance the modeling efficiency. Optimal Latin hypercube sampling technique (OLHS), orthogonal design and response surface model are then combined effectively to construct the approximate models between inputs and responses. Optimal design solutions are found successfully by applying the RBIS and MOPSO in the inner and outer loop respectively.

The remaining part of the paper is organized as follows. Section 5.2 studies the crashworthiness of the side door NPR impact beam by comparing to traditional side door impact beam and aluminum foam filled impact beam. The multi-objective reliability optimization process is detailed described in Sect. 5.3. Optimization results of the side door NPR impact beam are obtained in Sect. 5.4 and discussions are also conducted. Section 5.5 outlines the conclusions of the paper.

n of the side door impact beam plays an important role in determining the side impact crashworthiness of the vehicle. Detailed studies which focus on finding an optimum cross-section shape of the side impact beam were conducted by Li to improve the energy absorption character. Lee designed a thin-walled side door beam using quenchable boron steel and geometry optimization is also performed via creating new reinforcing ribs to promote the performances. Yoon developed a new one-body door impact beam by applying the advanced high strength steel to improve the impact energy absorbability and proper cross section shape of the door impact beam was also obtained in the studies. These studies improved the performances of side door impact beam significantly. However, it is challenging to make further improvement on the performances of the side door impact beam due to the limitation of space and cost. By combining the traditional side door impact beam and NPR structure, a novel side door NPR impact beam is first proposed in this work to make further improvement of the side impact crashworthiness under the limitation of space. The performances of the side door NPR impact beam will be detailed studied by comparing to the traditional side door impact beam and aluminum foam filled impact beam in Sect. 5.2.

Based on the composition of the side door NPR impact beam, it can be concluded that the performances of the parts are affected by the shape of the traditional side door impact beam, parameters of the NPR structures and material types, etc. To maximize the potential of the side door NPR impact beam, it is meaningful to find the optimal design to provide better passenger protection. Optimization-based design methods have been applied successfully in the past years by a lot of researchers to improve performances and reduce weight. However, the structure optimization studies are usually based on the assumption that all the variables and design parameters are determined exactly and there is no uncertainty. In the practical engineering industry, there are certainly errors, which are caused by human or natural factors, in the design and manufacture. The optimal design obtained by the deterministic optimization will become unreliable or even wrong when the constraints are violated easily. Thus, the uncertainly of the variables should be considered in the optimization to make the optimization results more reliable. In this work, the multi-objective design optimization of the side door NPR impact beam with considering reliability is performed to improve the performances. The parametric model of the NPR structure is first established to enhance the modeling efficiency. Optimal Latin hypercube sampling technique (OLHS), orthogonal design and response surface model are then combined effectively to construct the approximate models between inputs and responses. Optimal design solutions are found successfully by applying the RBIS and MOPSO in the inner and outer loop respectively.

The remaining part of the paper is organized as follows. Section 5.2 studies the crashworthiness of the side door NPR impact beam by comparing to traditional side door impact beam and aluminum foam filled impact beam. The multi-objective reliability optimization process is detailed described in Sect. 5.3. Optimization results of the side door NPR impact beam are obtained in Sect. 5.4 and discussions are also conducted. Section 5.5 outlines the conclusions of the paper.

5.2.2.2 Performances Studies of the Side Door NPR Impact Beam

To push the limit of space and improve the energy absorption capacity in side impact, a new side door NPR impact beam is first proposed in this work. The side door NPR impact beam is assembled by combining the NPR structure filled core and the traditional side door impact beam with cross section of rectangle. The FEM model of the side door NPR impact beam is first established, as shown in Fig. 5.22. 555,377 nodes and 407,568 elements are contained in the FEM model. The substrate material of the NPR structure filled core is aluminum, whose density is 2700 kg/m^3. The elastic modulus of aluminum is 70 GPa and the Poisson's ratio is 0.3. The material of the traditional side door impact beam is SAPH440, the density is 7850 kg/m^3, the elastic modulus is 210 GPa and the Poisson's ration is 0.3.

To study the performances, the side door NPR impact beam is fixed in the left front door of a minivan. Figure 5.23a is the automobile model crashed by a movable deformation counterguard in the side impact. Simulation and analysis are all conducted under the requirements of GB 20,071–2006 regulation. The total mass, which includes the mass of the movable deformation counterguard and crash dummy, is 2335 kg. The impact velocity is 50 km/h and the collision direction is perpendicular to the side of the car. The intrusion displacements of the measure points in the inner panel corresponding to thorax, abdomen and pelvis of the crash dummy are the most important indicators to evaluate the side impact crashworthiness, shown as

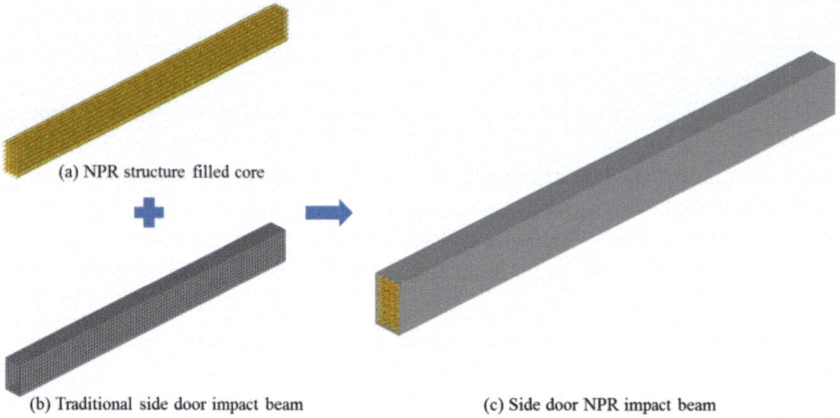

Fig. 5.22 FEM model of the side door NPR impact beam

Fig. 5.23b. Meanwhile, the intrusion velocities of the measure points in front door corresponding to crash dummy are another performance indicators.

Before performance simulation, the accuracy of the crash model is first validated by the analysis of energy change. The hourglass energy and interface energy of the collision are 0.39 and 1.36% respectively. Thus, it can be concluded that the collision model is accurate enough to be applied in the performance analysis. To study the crashworthiness performances and make comparisons, the traditional side door impact beam and the aluminum foam filled impact beam are also fixed respectively in the left front door of the minivan. It should be noted that the three different models are analyzed under the same conditions. The intrusion displacements and intrusion velocities of the three different models in the side impact are compared respectively in Figs. 5.24 and 5.25. Since the changing trend of intrusion displacements and

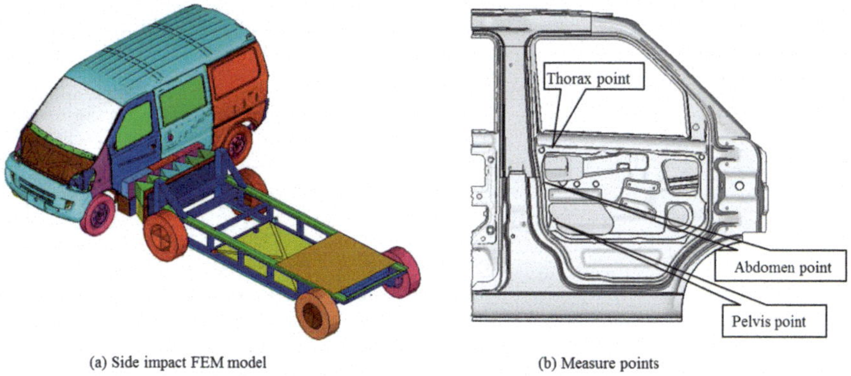

Fig. 5.23 Side impact FEM model and measure points of the vehicle

intrusion velocities in different measure points are almost the same, only the intrusion displacements and intrusion velocities of the measure point corresponding to thorax are listed owing to space constraints.

It can be seen from Fig. 5.24 that the model based on the side door NPR impact beam has the smallest peak intrusion displacement and the peak intrusion displacement of the model based on aluminum foam filled impact beam is smaller than that of the model based on traditional side door impact beam. Figure 5.25 shows that the intrusion velocity of model based on the side door NPR impact beam is smaller than those of the model based on traditional side door impact beam and aluminum foam

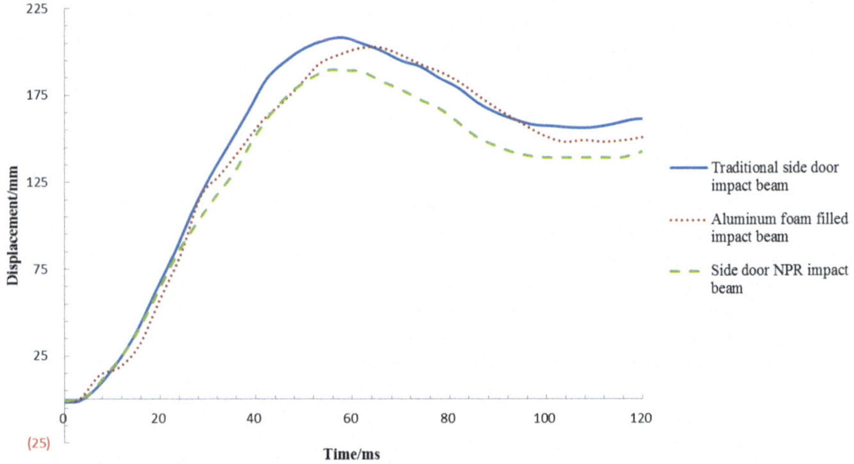

Fig. 5.24 Intrusion displacements of the measure points corresponding to thorax

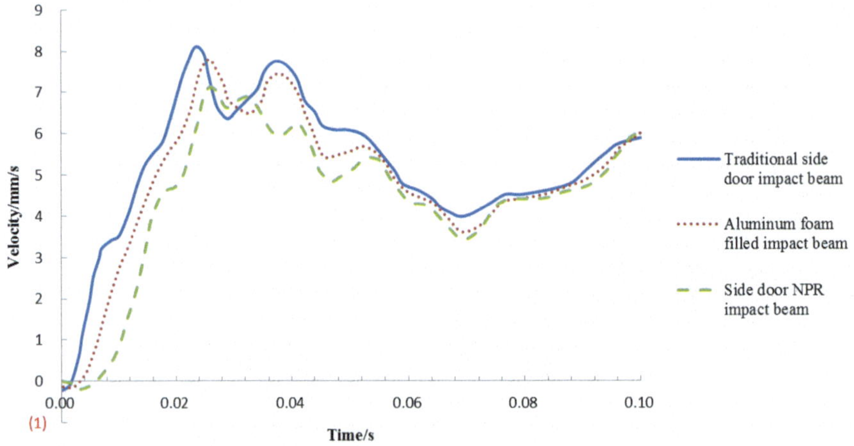

Fig. 5.25 Intrusion velocities of the measure points corresponding to thorax

Table 5.11 The maximum intrusion displacements and intrusion velocities of the three different models

Type of side door impact beam	Measure points corresponding to thorax		Measure points corresponding to abdomen		Measure points corresponding to pelvis	
	Dis./mm	Vel./m/s	Dis./mm	Vel./m/s	Dis./mm	Vel./m/s
Side door NPR impact beam	188.68	7.13	151.73	7.26	131.62	6.72
Aluminum foam filled impact beam	202.44	7.80	170.91	7.74	145.74	7.01
Traditional side door impact beam	208.28	8.10	177.63	8.02	156.41	7.25

filled impact beam during most of the collision process. The maximum intrusion displacements and intrusion velocities of the three different models are also listed in Table 5.11 to make the comparisons. It can be seen that smallest maximum intrusion displacements and intrusion velocities of the three different measure points can be obtained by the model based on the side door NPR impact beam. The model based on traditional side door impact beam has the largest maximum intrusion displacements and intrusion velocities. Therefore, it can be concluded that the side impact crashworthiness of the vehicle can be improved remarkably and good protection of passengers can be provided by applying the side door NPR impact beam in vehicle design.

5.2.2.3 Multi-objective Reliability Optimization Method

Although better side impact crashworthiness of the vehicle can be obtained by applying the side door NPR impact beam, the performances of the new parts are affected by a lot of parameters such as design parameters of the NPR structures, shape of the side door impact beam, material types, etc. To push the limit of the performances, a multi-objective reliability optimization method, which is established by combining the parametric model of the NPR structure, OLHS technique, RBIS and MOPSO, is proposed in this work to find the optimal design.

The NPR structure filled core of the side door NPR impact beam is assembled by repeating the NPR basic structure in three directions. Figure 5.26 is the enlargement of the NPR basic structure. It can be seen that l_1 and l_2 are the length of the long inclined cell wall and short inclined cell wall respectively, T_1 and T_2 are the thickness of the long inclined cell wall and short inclined cell wall respectively. H is the height of the NPR basic cell structure. θ is the angle between symmetry axis and long inclined cell wall, ϕ is the angle between symmetry axis and short inclined cell wall. W_1 is the width of the inclined cell wall, and it should be noted that the width of the long inclined cell wall and short inclined cell are the same. The shape and topology of the NPR structure will be changed when the design parameters are changed. By studying the relationships of the parameters, it can be known that l_1, θ, ϕ, T_1, T_2,

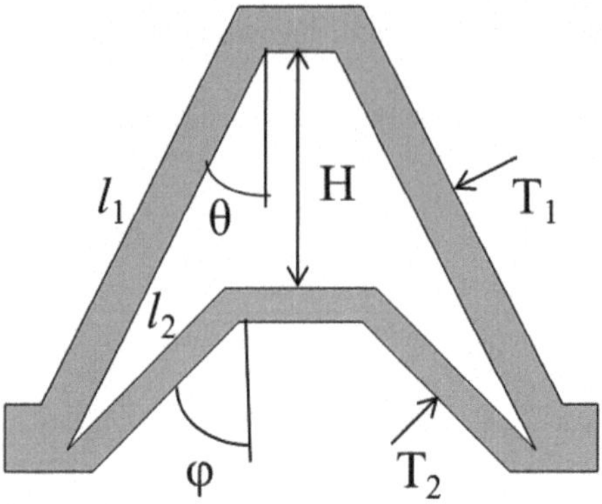

Fig. 5.26 Enlargement of the NPR basic cell structure

W_1 are the dominant parameters. To construct the FEM analysis model efficiently, the parametric model of the NPR structure is established based on the relationships of the parameters. When the design parameters are determined, the element will be generated adaptively and the models will be constructed automatically to be analyzed.

Based on the parametric model, the OLHS technique, orthogonal design and response surface model technique are combined to establish the surrogate models. To make the sample points distributed more uniformly, OLHS technique and orthogonal design are applied to conduct the design of experiment analysis. 4 sample points are generated in the outer array with the random parameters and 150 sample points are generated in the inner array with the design variables. Thus, 600 points will be generated to consider the uncertainty of noise factors by combining the points in the outer and inner array respectively. Response surface model is then applied to construct the surrogate models to reduce the expensive numerical simulations and improve the computational efficiency. After the surrogate models are established, three assessment methods, which are widely used in engineering, are applied to evaluate the accuracies. The squared correlation coefficient (R^2), relative maximum absolute error (RMAE) and relative average absolute error (RAAE) are given as follows.

$$R^2 = 1 - \frac{\sum_{i=1}^{q} (y_i - \hat{y}_i)^2}{\sum_{i=1}^{q} (y_i - \bar{y})^2} \tag{5.5}$$

$$RMAE = \frac{\max\{|y_i - \hat{y}_i|, \ldots, |y_i - \hat{y}_q|\}}{\sum_{i=1}^{q} |y_i - \bar{y}|/q} \tag{5.6}$$

$$RAAE = \frac{\sum_{i=1}^{q} |y_i - \hat{y}_i|}{\sum_{i=1}^{q} |y_i - \bar{y}|} \tag{5.7}$$

where y_i is the observed response value by the surrogate models, $\bar{y}$ is the mean value of y_i, $\hat{y}_i$ is the predicted value and q is the number of sampling points.

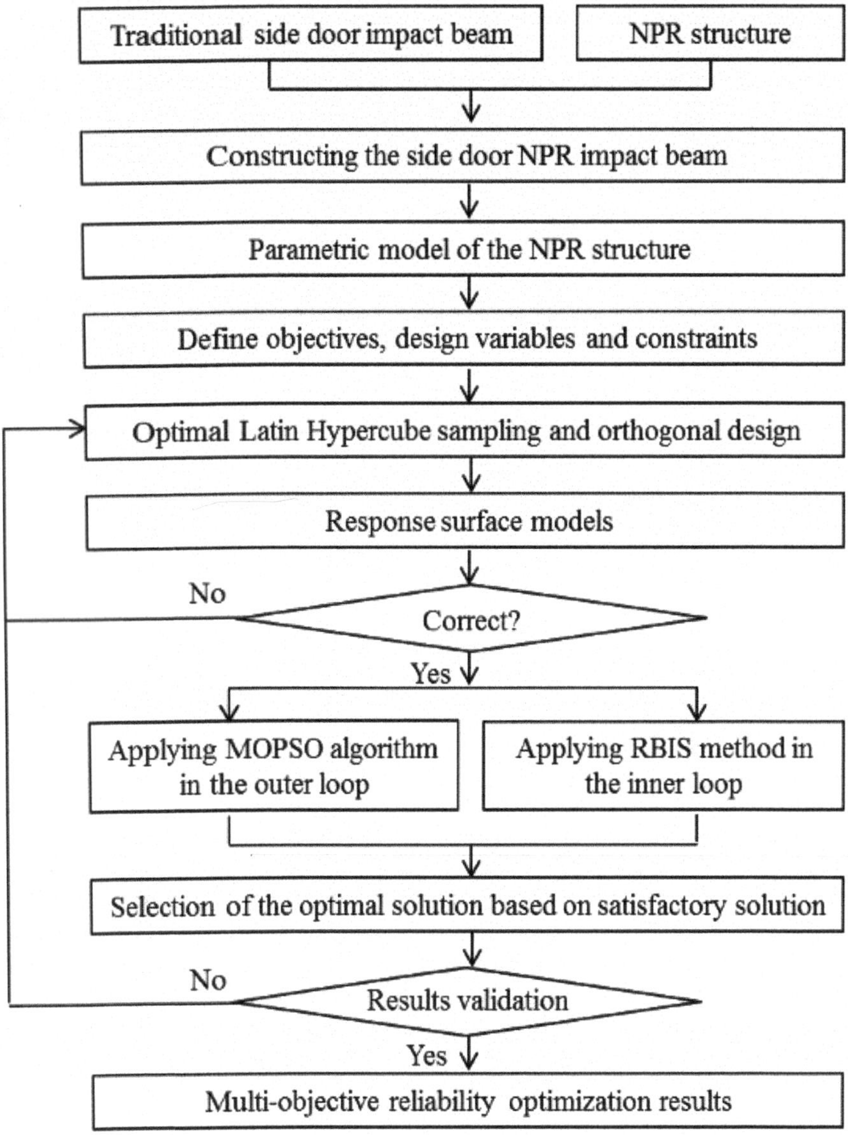

Fig. 5.27 Multi-objective reliability optimization process of the side door NPR impact beam

the random parameter, P^{rea} is the probability against joint chance constraint, R^{rea} is the designed reliability in the optimization.

After the response surface models are constructed, the three assessment methods are applied to evaluate the accuracies. The results of the accuracy assessment are listed in Table 5.13. It can be concluded that the surrogate models are accurate enough and can be applied in the optimization process to find the optimal solutions.

To perform the multi-objective reliability optimization design of the sid NPR impact beam, MOPSO, which has shown well-distributed Pareto front a convergence, is applied in the outer loop to conduct the design optimization RBIS is applied in the inner loop to carry out the reliability analysis. The calcula P_f, which is the general form of probability of failure, can be converted to conc sampling outside the β-sphere, given as follows.

$$P_f = \frac{N_{fail}}{N_{sim}}\left[1 - \chi_n^2(\beta^2)\right]$$

where N_{fail} is the number of failed sample points, N_{sim} is the number of sample outside the β-sphere and χ_n^2 is the chi-square distribution function. The d description of the RBIS method can be found in the references provided by H and Grooteman. The multi-objective reliability optimization process of the sid NPR impact beam can be given as Fig. 5.27.

5.2.2.4 Optimization Results and Discussions

To improve the side impact crashworthiness of the vehicle, the intrusion dis ments of the measure points should be as lower as possible and the intrusion vel should be controlled in a safe range. The weight of the vehicle has great effect fuel consumption and emission, it should be as lighter as possible in the prem good performances. In this work, the intrusion displacements of the measure and the mass of the side door NPR impact beam are selected as the design tives, the intrusion velocities are selected as the constraints. The information design variables and random parameters for the side door NPR impact bea summarized in Table 5.12. The width (W_2) and height (H_2) of the cross sectio selected as the deterministic design variables. The design parameters of the structure (l_1, θ, ϕ, T_1, T_2, W_1) are selected as the random design variables, the density (ρ), Young's modulus (E) and yielding stress (σ) of the steel are cl as random parameters in this work. It should be noted that all the design vari and random parameters are assumed to be independent and disturbed normally desired coefficient of variation is set as 0.03 in this work.

Based on the above analysis, the corresponding multi-objective reliability d optimization model of the side door NPR impact beam can be expressed as foll

$$\begin{cases} \underset{(x_d, \mu_x)}{\text{Min}} \quad \left\{D_{ins}\left(x_d, \mu_{x_r}, \mu_p\right), M\left(x_d, \mu_{x_r}, \mu_p\right)\right\} \\ \text{Subject to: } \mathrm{P}^{rea}\left(\left(V_{Ins}(x_d, x_r, p) \le 7.13\right)\right) \ge R^{rea} \\ x_d^L \le x_d \le x_d^U, \mu_{x_r}^L \le \mu_{x_r}, \le \mu_{x_r}^U \end{cases}$$

where D_{ins} is the intrusion displacement, M is the mass of the side door NPR im beam, x_d is the deterministic design variable, x_r is the random design variable, displacement is obtained, the larger mass is needed with the same reliability level and a smaller mass will lead to larger intrusion displacement. To select the most satisfactory optimal solution, the satisfactory function is applied in this work. shown as follows.

$$S = \frac{D_{ins} - (D_{ins})_{\min}}{(D_{ins})_{\max} - (D_{ins})_{\min}} + \frac{M - M_{\min}}{M_{\max} - M_{\min}} \tag{5.10}$$

where $(D_{ins})_{\max}$ and $(D_{ins})_{\min}$ are the maximum and minimum value of the intrusion displacement among the Pareto optimal front. $M_{\max}$ and $M_{\min}$ are the maximum and minimum value of mass among the Pareto optimal front. The lower is the value of S, the better is the optimal solution.

By applying the satisfactory function, the optimal design is selected from the Pareto optimal front which is obtained under the 99% reliability level, given in Table 5.15. The original design and the optimal design obtained by the deterministic optimization are also listed in the table to make the comparisons. The comparisons of performances with the multi-objective reliability and multi-objective deterministic optimization are given in Table 5.16. FEM analyses based on the optimal solutions are also conducted in this work to verify the design results.

The verification by the FEM analysis shows that there is little error between results of response surface models and traditional models. Therefore, the optimal

Table 5.15 Design variables obtained by reliability and deterministic optimization

Design variable	Original design	Value obtained by deterministic optimization	Value obtained by reliability optimization
W_2/mm	48.00	53.27	51.95
H_2/mm	90.00	82.34	84.67
l_1/mm	10.00	9.34	9.07
θ/rad	0.32	0.37	0.31
φ/rad	0.48	0.51	0.43
T_1/mm	0.90	0.82	0.79
T_2/mm	0.90	0.94	1.03
W_1/mm	1.20	1.31	1.38

Table 5.16 Comparison of performances with the reliability and deterministic optimization

	D_{ins}/mm	M/Kg	V_{Ins}/m/s
Original design	188.68	4.59	7.13
Multi-objective deterministic optimization	172.63	4.29	7.01
Verified deterministic optimization by FEM	173.70	4.24	6.99
Multi-objective reliability optimization	174.27	4.33	7.10
Verified reliability optimization by FEM	175.46	4.30	7.12

results obtained from the reliability and deterministic optimizations are reliable. It can be seen from the Tables 5.15 and 5.16 that the performances of the side door NPR impact beam are further improved successfully by the multi-objective reliability and multi-objective deterministic optimization. The intrusion displacement and intrusion velocity of the optimal design obtained from the multi-objective deterministic optimization are less than those of the multi-objective reliability optimization. However, the reliability level of the optimal design obtained from the multi-objective reliability optimization is much higher than those of the multi-objective deterministic optimization. Thus, it can be concluded that more reliable optimal solution can be obtained by the multi-objective reliability optimization at the expense of performance. The results also show that better performances of the side door NPR impact beam can be obtained by the multi-objective reliability optimization with high reliability by comparing to the original side door NPR impact beam.

5.2.2.5 Conclusions

(1) A novel side door NPR impact beam is first proposed in this work to improve the vehicle side impact crashworthiness by combining the traditional side door impact beam and NPR structure. The performances of the side door NPR impact beam are detailed studied by comparing to traditional side door impact beam and aluminum foam filled impact beam. The results show that the intrusion displacement and intrusion velocity of the vehicle based on the side door NPR impact beam are less than those of the vehicle based on traditional side door impact beam and aluminum foam filled impact beam. Better side impact crashworthiness can be provided by the side door NPR impact beam to make better protection of the passenger.

(2) The multi-objective reliability design optimization of the side door NPR impact beam is conducted successfully in this work to make further improvement of the performances with considering reliability. Parametric model of the NPR structure is first established to improve the modeling efficiency. Surrogate models are then built by combining the Latin hypercube sampling technique, orthogonal design and response surface model. The multi-objective reliability optimal solutions are found successfully by applying the MOPSO algorithm in the outer loop and RBIS method in the inner loop respectively. Multi-objective optimization with different reliability level and deterministic optimization are also performed respectively to make a detailed study. The results show that better performances of the side door NPR impact beam can be obtained by optimizing the design parameters. The side impact crashworthiness of the vehicle is further improved by the multi-objective reliability optimization and better protection of passenger is provided. This work also serves a good example for other performances improvement of automobile.

5.2.3 Application of Negative Poisson's Ratio Structure in Energy-Absorbing Structure for Pedestrian Protection

5.2.3.1 Introduction

According to the World Health Organization, 22% of all automobile fatalities are pedestrians. With the development of automobile safety technology, pedestrian protection has attracted increasing emphasis. Because lower leg injuries account for the largest proportion of all vehicular injuries involving pedestrians, developing measures for enhancing pedestrian lower leg protection represents the most meaningful approach for improving pedestrian protection. The lower leg generally makes first contact with the front-end structure of a vehicle in vehicular accidents involving pedestrians. As such, the degree of injury is determined by the energy absorption capacity of the front-end structure. Therefore, generally speaking, increasing the energy absorption capacity of the front-end structure of vehicles will reduce the impact of collision, which includes features such as tibia acceleration, knee bending angle, and knee shear displacement. Correspondingly, optimizing the energy absorption capacity of vehicle front-end structures has attracted considerable interest. Lee et al. conducted a variance analysis of pedestrian lower leg injuries due to collision with the front-end structure of vehicles. Based on the analysis, the front-end structure was optimized, and the risk of pedestrian leg injury was minimized successfully. Lv et al. effectively enhanced pedestrian lower leg protection by optimizing the parameters of a vehicle front-end structure using an ensemble of meta-models and a multi-objective optimization algorithm. Zhang et al. added energy absorption foam and an extra bumper according to optimizations of the cross section of energy absorption foam to enhance pedestrian lower leg protection. While the above studies were able to successfully enhance pedestrian lower leg protection by optimizing the performance of front-end structural features, the limited design space available in the front end of vehicles presents significant challenges toward making further improvements in pedestrian lower-leg protection using conventional front-end structures. Therefore, the present work proposes a novel negative Poisson's ratio (NPR) energy absorption structure with high energy absorption capacity to further improve pedestrian lower-leg protection.

The energy absorption performance of the proposed NPR energy absorption front-end structure is strongly dependent on the selected design parameters, and the performance can vary widely with changing parameters. Therefore, the design parameters of the NPR energy absorption structure should be optimized to maximize the energy absorption capacity of the front-end structure. Structural optimization employing the finite element method (FEM) and advanced optimization algorithms has developed apace with rapidly advancing computer hardware and software technology. To meet national safety standards of pedestrian lower leg protection, structural optimization has also been successfully applied by numerous researchers for improving pedestrian protection measures. Han and Lee conducted multi-objective optimization by combining sequential quadratic programming and an approximation model to improve the pedestrian lower leg protection provided by a vehicle font-end structure. Mohapatra conducted rapid optimization design for automotive bumper energy absorbers by applying a morphing technique. Hui successfully minimized tibia acceleration and knee shearing displacement owing to lower-leg contact with a vehicle front-end by combining experimental design, meta-models, and a multi-objective optimization algorithm. Sun et al. optimized the parameters of a bumper system using an orthogonal test method and a main-effect analytical method to improve pedestrian protection. The present work optimizes the energy absorption capacity of the proposed NPR front-end structure by conducting multi-objective optimization with consideration for the perturbation induced by parameter uncertainties. A parametric model of the NPR energy absorption front-end structure is first built based on the relationships between design parameters, and approximation models between inputs and responses are set up by combining an optimal Latin hypercube sampling (OLHS) technique and the dual response surface method. The optimal NPR front-end structure design is then successfully obtained by combining a multi-objective particle swarm optimization (MOPSO) algorithm and six sigma criteria. Finally, the optimization results are validated by comparisons with the results of multi-objective deterministic optimization.

The remainder of this paper is organized as follows. The NPR front-end structure is introduced in Sect. 5.2, and the separate types of performance of the proposed NPR front-end structure are evaluated by comparison with the performances of conventional energy absorption structures. Section 5.3 introduces the multi-objective robust design optimization process. Section 5.4 introduces the details of optimization for the NPR energy absorption structure. Multi-objective robust and deterministic optimizations are respectively applied to the NPR front-end structure in Sect. 5.5, and comparisons are conducted. Section 5.6 outlines the conclusions of this work.

5.2.3.2 Performance Studies of the NPR Energy Absorption Structure

The proposed NPR energy absorption structure is assembled from a basic NPR cell structure repeated in three directions, as illustrated in Fig. 5.1a. Like the conventional energy absorption structures, the NPR energy absorption structure is composed of foam and the base material of the foam is polyurethane. The elastic modulus of the foam is 5.0 MPa, Poisson's ratio is 0.20, and the density is 30 kg/m^3, the yield strength of the foam which used in this work is 0.22 Mpa. Here, we note that 3D printing, which is mainly applied as rapid prototyping and layered manufacturing, is a potential method for efficiently manufacturing the proposed NPR structure. Detailed discussion of the 3D printing and additive manufacturing can be found elsewhere. It should be noted that only the hardening effect and strain rate effect of the steel are considered, those of the foam are not taken in account due to the material properties and condition restriction. The NPR energy absorption structure is installed between the bumper and anti-collision beam, as shown in Fig. 5.29b. Because the pedestrian protection performance is mainly determined by the front-end structure of the vehicle, only the model of the overall front-end structure shown in Fig. 5.29 is applied in the FEM computations to provide a good balance between efficiency and accuracy. The front-end model contains 1,015,753 elements and 826,626 nodes. It should be noted that all statistical data employed in the modeling work conducted in the present study were obtained from material tests conducted by the supplier. 57th material in the material library of software package LS-DYNA is applied as the constitutive model for NPR structure. Details regarding the constitutive model can be found elsewhere.

Regarding the modeling of the human leg employed in the present study, we note that a study conducted by Matsui showed that the differences between the knee injury of the TRL leg-form and the knee injury of the Flex-PLI leg-form are not remarkable. Therefore, the TRL leg-form was applied in this work. The performance of the NPR energy absorption structure in conjunction with the Flex-PLI leg-form will be discussed in a subsequent study to bring the results closer to the latest standards. We conducted Transport Research Laboratory (TRL) impact tests based on the GB/T 24,550–2009 standard, where collisions between the leg-form and the front-end structure occur at the speed of 40 km/h. We considered tibia acceleration, knee bending angle, and knee shear displacement as the performance indicators of pedestrian lower leg protection. To assess the performance indicators objectively, three measurement points were considered, as shown in Fig. 5.30. First, we compared the performance indicators obtained for three different models at all three measurement points. The model based on a conventional energy absorption structure, the model based on an optimized energy absorption structure, and the model based on the proposed NPR energy absorption structure are herein denoted as Model 1, Model 2, and Model 3, respectively to make the description more concise. The optimized energy absorption structure was constructed by optimizing the shape of the conventional energy absorption structure with the addition of five holes for mass reduction. All models were computed and analyzed under equivalent conditions.

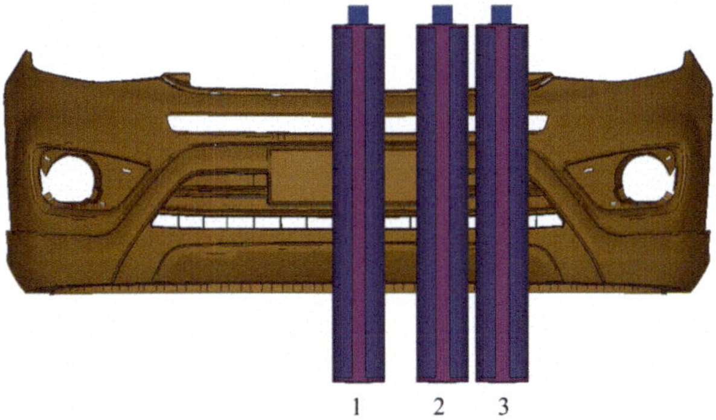

Fig. 5.30 The three measurement points on the vehicle front-end structure employed for testing

The performance indicators obtained from TRL impact tests as a function of time for the three different models at all three measurement points are compared in Fig. 5.31. The masses of the energy absorption structures and the peak values obtained are also listed in Table 5.1. It can be seen from Fig. 5.31 and Table 5.17 that the performance indicators for Model 2 and Model 3 are less than those obtained from Model 1, which indicates that the pedestrian lower leg protection performance of the vehicle is promoted by both optimizing the conventional energy absorption structure and by employing the NPR energy absorption structure. In addition, from Table 5.17 we note that the values of the performance indicators for Model 3 are generally less than those for Model 2. Thus, it can be concluded that the NPR energy absorption structure provides the best pedestrian lower leg protection among the structures considered. We also note from Table 5.17 that the best pedestrian lower leg protection is provided by the proposed NPR energy absorption structure with the smallest mass.

5.2.3.3 Multi-objective Robust Optimization Process of the NPR Energy Absorption Structure

The multi-objective robust optimization process applied to the NPR energy absorption structure is presented in Fig. 5.32. Based on the parameter relationships of the NPR basic cell structure, a parametric model of the NPR energy absorption

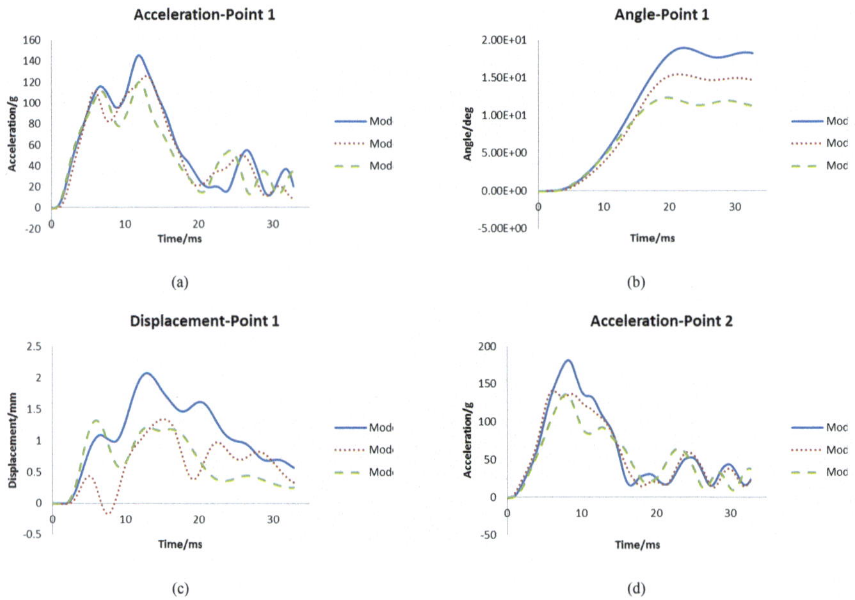

Fig. 5.31 Performance comparisons of the three different models

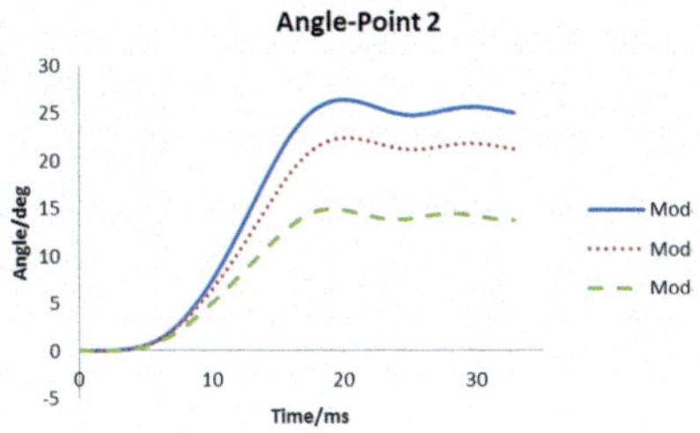

(e)

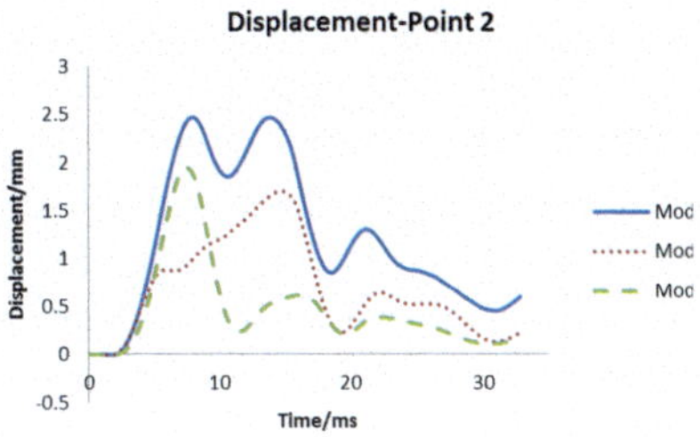

(f)

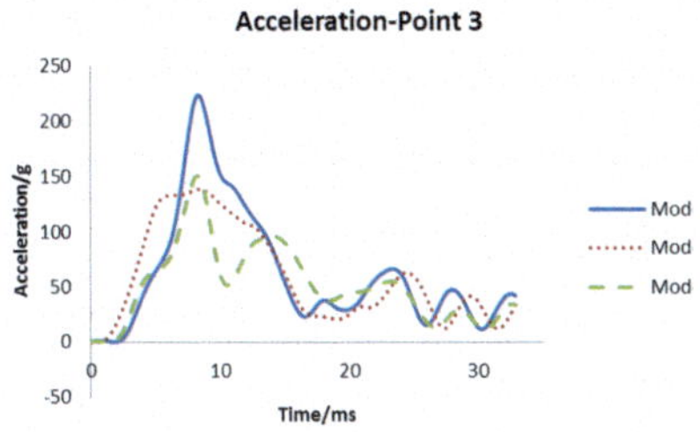

(g)

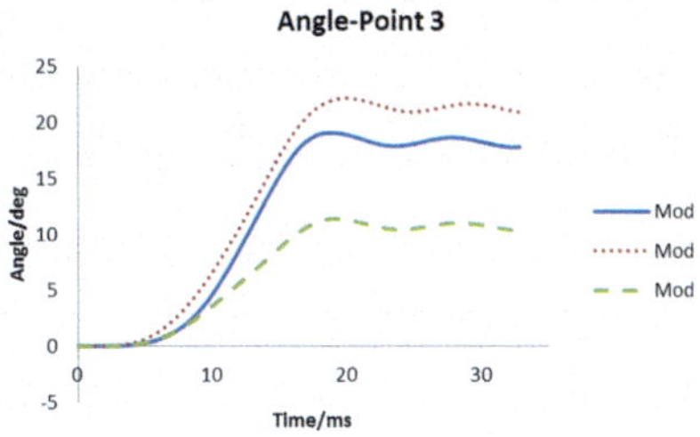

(h)

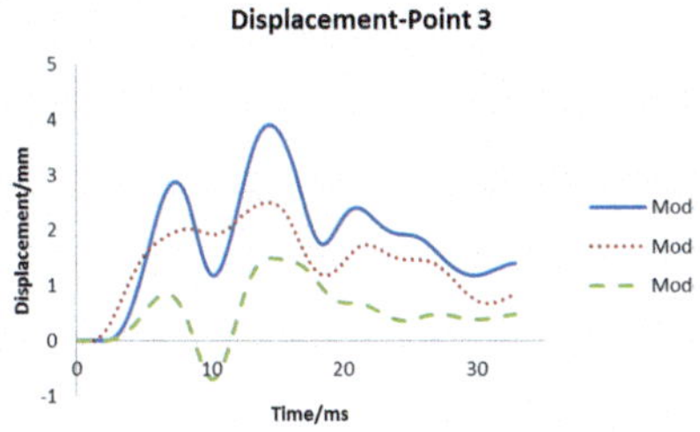

(i)

Fig. 5.31 (continued)

Table 5.17 The mass of the energy absorption structure and the peak values for the three different models

	Mass (kg)	Tibia acceleration (g)			Knee bending angle (°)			Shear displacement (mm)		
		P1	P2	P3	P1	P2	P3	P1	P2	P3
Model 1	0.4586	145.35	181.67	223.52	19.01	26.46	19.15	2.15	2.54	3.98
Model 2	0.2627	125.86	141.18	138.92	15.54	22.53	22.23	1.38	1.77	2.56
Model 3	0.2417	120.13	136.46	150.72	12.47	15.09	11.49	1.34	1.99	1.57

structure is first built to improve the modeling efficiency and accuracy. The OLHS technique and dual response surface method are combined to establish surrogate models for representing pedestrian collision to promote computational efficiency. The multi-objective robust optimization is then set up based on six sigma criteria. The MOPSO algorithm is applied to obtain the optimal frontal design solutions and optimal solutions are finally obtained by applying the minimum distance selection method.

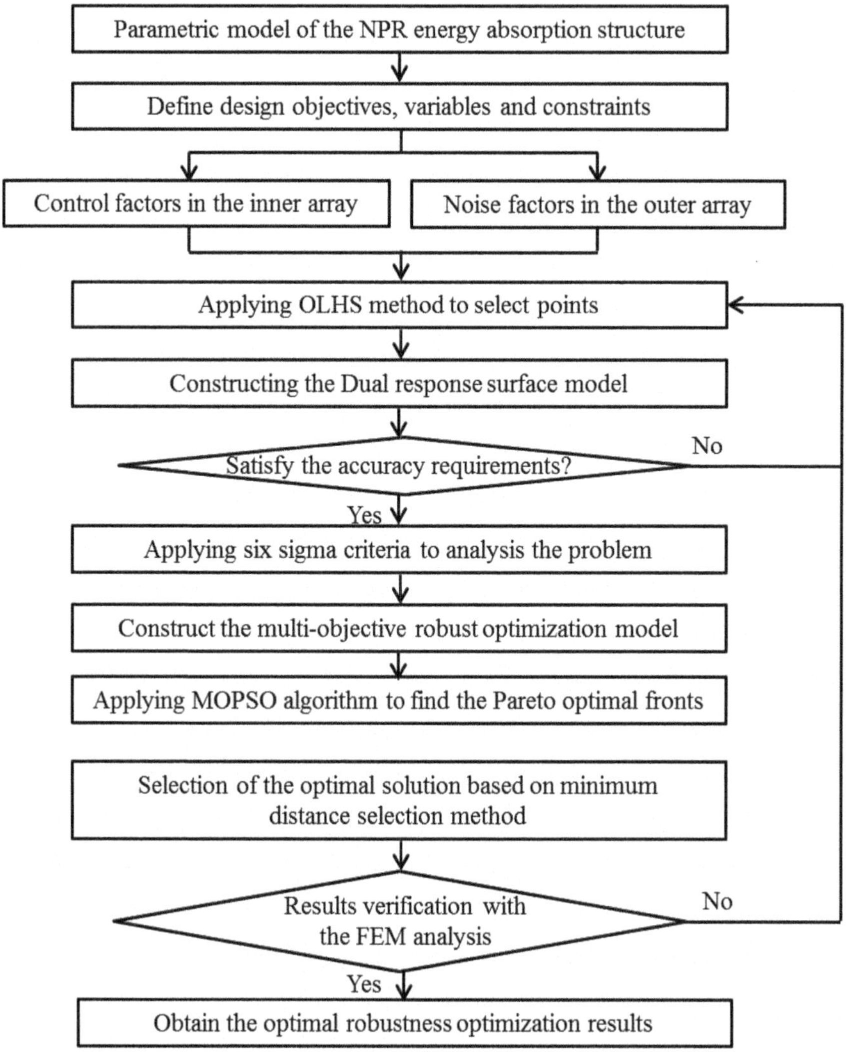

Fig. 5.32 Multi-objective robust optimization process applied to the NPR energy absorption structure

To construct the surrogate models, orthogonal design is first applied to generate the sample points. The accuracy of the approximate response surface models can be improved by employing a more uniform distribution of sample points. Therefore, the OLHS technique, which provides a more uniform level distribution of sample points by optimizing the level order in each column, is applied in this work to generate the sample points. Detailed descriptions of the OLHS technique can be obtained elsewhere. To construct the dual response surface models, all design parameters are arranged into two independent arrays to generate a cross product array. Noise factors are placed in the outer array to consider the parameter uncertainties, and the control factors are placed in the inner array. Details regarding the inner and outer arrays can be obtained elsewhere. Unlike the conventional response surface model technique, the mean response surface functions of the performance indicators and standard deviations are fit separately in the dual response surface method. The dual response surface functions are then applied to conduct the multi-objective robust optimization in combination with six sigma criteria. The six sigma robust optimization design method, which was first proposed and applied by Motorola Inc., is mainly applied to reduce the sensitivity and enhance the robustness of the optimization process. Specifically, application of six sigma criteria in the optimization design facilitates the optimization of the performance indicators by considering the perturbation induced by parameter uncertainties. Based on the fundamental principle of six sigma criteria, we know that the mean values of the performance indicators should be as close as possible to their target values, and their variances should be as low as possible in the optimization process.

The overall performance of the NPR energy absorption structure is mainly determined by the material and design parameters of the basic cell structure in three directions. The NPR basic cell structure is illustrated in Fig. 5.33, where l_1 and T_1 and l_2 and T_2 are the lengths and thicknesses of the long inclined cell wall and short inclined cell wall, respectively, W is the width of the long and short inclined cell walls, θ is the angle between the long inclined cell wall and the vertical axis, and ϕ is the angle between the short inclined cell wall and the vertical axis. The relationships between the parameters of the NPR basic cell structure are expressed in Eq. (5.11) according to its geometry. It can therefore be concluded that the parameters cannot be arbitrarily changed. In addition, it should be noted that the top horizontal segments of the long and short cell walls have the same length. The parameters of the NPR structure not only determine the shape and topology of the NPR energy absorption structure, but also have a significant impact on its performance and mass. Therefore, a parametric model of the NPR energy absorption structure is established based on Eq. (5.11) using l_1, l_2, θ, W, T_1, and T_2 as the modeling parameters. It should be noted that ϕ can be obtained from Eq. (5.11), and is therefore not selected as a design parameter. The FEM nodes are then calculated from the determined modeling parameters, and the model elements are generated automatically and adaptively.

$$\cos\theta = \frac{l_1^2 + (l_1\cos\theta - l_2\cos\varphi)^2 - l_2^2}{2l_1(l_1\cos\theta - l_2\cos\varphi)} \tag{5.11}$$

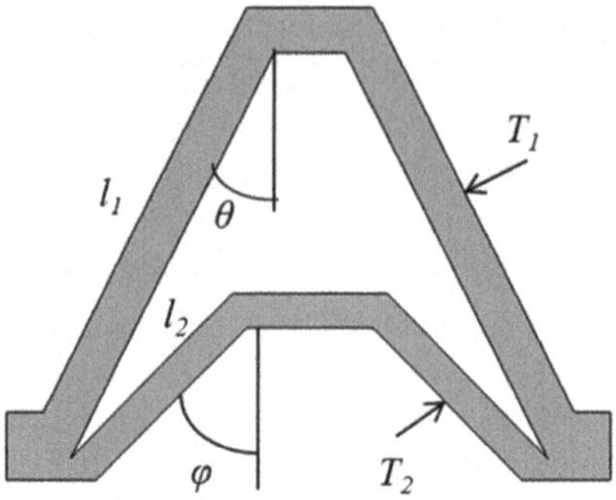

Fig. 5.33 Design parameters of the basic NPR cell structure

To improve the computational efficiency of the optimization, surrogate models of the pedestrian collision model are constructed to replace the FEM models in the calculation. Because only a model of the mean response value without considering variance can be set up in the conventional response surface model technique, the dual response surface model is applied in this work to build models of the mean response values and standard deviations simultaneously. Further details regarding the dual response surface model can be found elsewhere. The models are given as follows.

$$\begin{cases} y_\mu = \tilde{y}_\mu + \varepsilon_\mu = \sum_{i=1}^{I} m_i \phi_i(X) + \varepsilon_\mu \\ y_\sigma = \tilde{y}_\sigma + \varepsilon_\sigma = \sum_{j=1}^{J} n_j \varphi_j(X) + \varepsilon_\sigma \end{cases} \tag{5.12}$$

Here, X represents the design parameters, y_μ and $\tilde{y}_\mu$ are the mean values of the responses respectively obtained from FEM and response surface models, y_σ and $\tilde{y}_\sigma$ are the standard deviations of the responses respectively obtained from FEA and response surface models, ε_μ and ε_σ are the respective modeling errors of the mean and standard deviations of the responses, and m_i and n_j are the unknown coefficients corresponding to the basis functions φ_i (X) and ϕ_j (X), respectively.

Because it is difficult to evaluate the accuracy of the surrogate models based on only a single means of evaluation, the squared correlation coefficient (R^2) and relative average absolute error (RAAE) are calculated to verify the accuracies of the established response surface models. The values of R^2 and RAAE are respectively given as follows.

$$R^2 = 1 - \frac{\sum_{i=1}^{k} (y_i - \tilde{y}_i)^2}{\sum_{i=1}^{k} (y_i - \overline{y})^2} \tag{5.13}$$

$$RAAE = \frac{\sum_{i=1}^{k} |y_i - \tilde{y}_i|}{\sum_{i=1}^{k} |y_i - \overline{y}|} \tag{5.14}$$

Here, k is the number of sampling points, $\tilde{y}_i$ is the predicted value of the responses obtained from the dual response surface models, y_i is the result of FEM and $\bar{y}$ is the mean value of y_i. Further details regarding R^2 and RAAE can be found in the cited references. Generally speaking, the accuracies of the response surface models increase with decreasing RAAE and as R^2 approaches 1. The accuracy requirement can be met with an RAAE value greater than 0.1 for this crashworthiness problem.

To consider the perturbation induced by parameter uncertainties, six sigma criteria are applied to construct the multi-objective robust optimization. For the two objectives, the robust optimization can be expressed as follows.

$$\begin{cases} \min\{f_1(Y_{\mu 1}(x), Y_{\sigma 1}(x)), f_2(Y_{\mu 2}(x), Y_{\sigma 2}(x))\} \\ s.t.\ g_{\mu j}(x) + 6g_{\sigma j}(x) \le 0 \\ x_L + 6x_\sigma \le x_\mu \le x_U - 6x_\sigma \end{cases} \tag{5.15}$$

Here, $Y_{\mu 1}(x)$ and $Y_{\mu 2}(x)$ represent the respective mean values of the two objectives, $Y_{\sigma 1}(x)$ and $Y_{\sigma 2}(x)$ are the respective standard deviations of the two objectives, $g_{\mu j}(x)$ and x_μ are the respective mean values of the constraints and design variables, $g_{\sigma j}(x)$ and x_σ denote the respective standard deviations of the constraints and design parameters, and x_U and x_L are the upper and lower bounds of the design variables, respectively. Since multiple different design objectives need to be considered simultaneously, the MOPSO algorithm and minimum distance selection method are then applied to the established multi-objective robust optimization for obtaining an optimal design solution. It should be noted that the performance indicators of the three different models are compared only at Point 3 to reduce the computational cost.

5.2.3.4 Multi-objective Robust Optimization Results and Discussion

According to the analysis of the performance indicators conducted in Sect. 5.2, meeting the requirements for tibia acceleration and knee bending angle is challenging, whereas the requirements for knee shear displacement are more easily satisfied for the front-end model considered; thus, tibia acceleration and knee bending angle were selected as optimization objectives and knee shear displacement was selected as an optimization constraint. The vehicle mass has a significant influence on fuel consumption and emissions; thus, the mass of the NPR energy absorption structure and anti-collision beam were also selected as constraints, and must be no greater than those of the original model. In addition, the thickness of the anti-collision beam (T_3) has an important impact on the performance, and was also selected as a design parameter. The anti-collision beam is composed of steel. Because the performance of steel is affected by the manufacturing environment and uncertainties in the material properties are also a factor, the Young's modulus (E), yielding stress (σ), and density (ρ) of the steel were selected as random parameters with fluctuations in the ranges of $E = (195, 205\ \text{GPa})$, $\sigma = (210, 220\ \text{GPa})$, and $\rho = (7600, 8000\ \text{kg/m}^3)$. Based on the six sigma criteria, the multi-objective robust optimization of the vehicle front-end structure can be established as follows.

$$\begin{cases} \min: \ (f_1, f_2) \\ f_1 = \lambda Acc_\mu^2 + (1-\lambda) Acc_\sigma^2 \\ f_2 = \lambda Angle_\mu^2 + (1-\lambda) Angle_\sigma^2 \\ s.t. \ Disp_\mu + 6 Disp_\sigma \leq 2.0 \\ m \leq m_{original} \\ D^L + 6D_\sigma \leq D_\mu \leq D^U - 6D_\sigma \end{cases}. \tag{5.16}$$

Here, D is a matrix of the design parameters and λ is a weighting factor for emphasizing either the mean or standard deviation objectives. It should be noted that all design parameters of the structure in this work were assumed to be normally distributed, and the standard deviations are [0.01, 0.01, 0.01]. In addition, λ has an important effect on the robustness of the optimization. It is hard to clearly point out what value is the optimum value for λ. Generally speaking, λ should be selected based on engineering practice and design requirements. Further details regarding λ are presented elsewhere. In the present work, λ was set as 0.1.

It should be noted that OLHS was conducted first without considering the constraints, and the initial sample points were selected based on the basic requirements that l_1 should be greater than l_2 and ϕ greater than θ. Based on the desired accuracy, efficiency, and the number of variables, 200 sample points were employed in the inner array and 4 sample points in the outer array. The simulations were repeated at every single sample point in the outer array to consider the variability caused by the uncertainties in the noise factors; thus, a total of 800 simulations were performed in this step to generate the surrogate models. All the simulations were conducted in the computer support group, which includes 32 computers, each with an 8 core CPU. The dual response model in quadratic polynomials was then applied to construct the surrogate models of the mass and all three performance indicators. Once the approximation models were constructed, the values of R^2 and RAAE were calculated to verify the accuracies of the models. The results are listed in Table 5.18. The verification results show that the approximation models are sufficiently accurate to be applied in the subsequent optimization process. The MOPSO algorithm was then applied to obtain the optimal design solutions. The parameters employed in the MOPSO algorithm are listed in Table 5.19.

The optimization results were obtained after 83 iterations of the MPOSO algorithm with the approximation models, and the Pareto optimal fronts of the multi-objective robust optimization are shown as Fig. 5.34. The optimization results

Table 5.18 Accuracies of the different approximation models

	Acc_μ	Acc_σ	$Angle_\mu$	$Angle_\sigma$	$Disp_\mu$	$Disp_\sigma$
R^2	0.9927	0.9943	0.9949	0.9951	0.9934	0.9953
RAAE	0.0271	0.0239	0.0217	0.0314	0.0277	0.0284

Table 5.19 Parameters set in the MOPSO algorithm

MOPSO parameters	Value
Population size	120
External archive	120
Inertial weight	0.65
Global learning coefficient	1.50
Personal learning coefficient	1.50

obtained using multi-objective deterministic optimization without considering the perturbation induced by parameter uncertainties are also given in Fig. 5.34.

It can be seen from Fig. 5.34 that the tibia acceleration and knee bending angle are reduced by optimizing the design parameters of the NPR energy absorption structure and anti-collision beam using both optimization methods. Comparison of the results of the two different optimization methods indicates that the performance associated with the multi-objective deterministic optimization results is greater than the performance associated with the multi-objective robust optimization results. However, because the perturbations induced by parameter uncertainties are considered in the multi-objective robust optimization, we note from the standard derivations of the objective functions for the two different optimizations listed in Table 5.21 that the standard derivations of the objective functions in the multi-objective robust optimization are less than those of the multi-objective deterministic optimization,

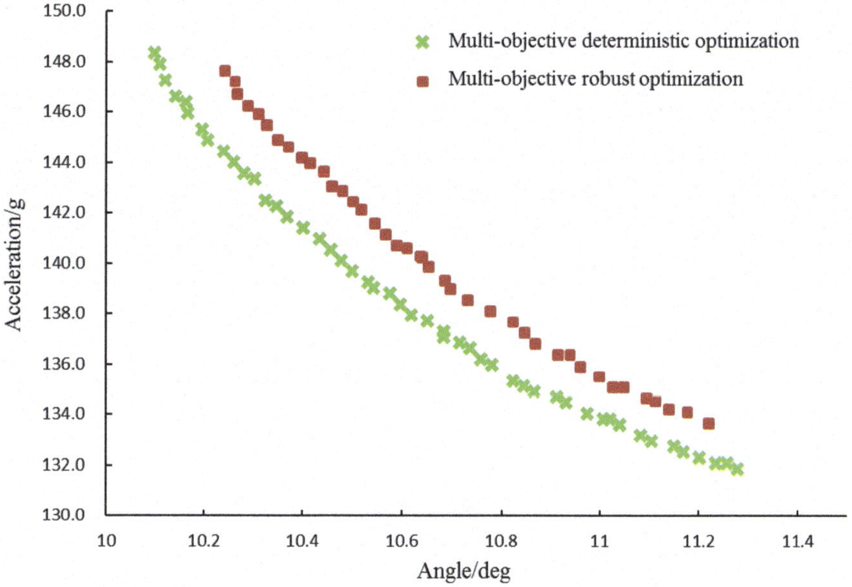

Fig. 5.34 Pareto optimal fronts of the two multi-objective optimization methods

which means that the robustness of the optimization is greater. This makes the multi-objective robust optimization results more suitable for application in practical vehicle design.

Because different optimization solutions are provided by the two different optimization methods, the minimum distance selection method was applied to select the optimum design. The evaluation criterion of the minimum distance selection method is given as follows.

$$\min D_{is} = \left(\sum_{i=1}^{n} (f_{ji} - \min(f_i(x)))^2 \right)^{\frac{1}{2}} \tag{5.17}$$

Here, D_{is} is the distance between the most satisfactory point and the ideal point, where the ideal point is determined by the optimal values of each individual optimization objective, n is the number of objectives, f_{ji} is the ith objective in the jth Pareto solution, and $min(f_i(x))$ is the minimum objective $f_i(x)$ among all the Pareto solutions. Generally speaking, the ideal point cannot be obtained due to conflicts between objectives. By applying the minimum distance selection method, two optimal design solutions were selected respectively from the Pareto solutions of the multi-objective robust optimization and multi-objective deterministic optimization. To provide for comparisons of the different optimization results, the design parameters and performances of the two optimal solutions are listed in Tables 5.20 and 5.21, respectively.

To verify the optimization results, FEM models based on the optimal design variables obtained from the two optimization methods were also constructed and analyzed. The analysis results are listed in Table 5.5. It can be seen that the results of the FEM models deviate little from the obtained optimal results. Thus, it can be concluded that the optimal results obtained by the different optimization algorithms have sufficient accuracies. Comparison with the FEM model results of the original design indicate that the performance of the NPR energy absorption structure was significantly enhanced by the two different optimization algorithms, resulting in better pedestrian lower leg protection. Although the performance obtained by the

Table 5.20 Optimal design variables obtained from the deterministic and robust optimization methods

Design parameter	Original value	Value obtained by robust optimization	Value obtained by deterministic optimization
l_l (rad)	20.29	19.34	18.74
l_2 (rad)	10.17	12.34	11.76
θ (rad)	0.52	0.48	0.55
W (mm)	4.00	4.43	4.76
T_1 (mm)	0.85	0.77	0.72
T_2 (mm)	0.85	0.89	0.96
T_3 (mm)	1.20	1.03	0.96

Table 5.21 Performances of the optimal solutions obtained by deterministic and robust optimization methods

	Acc_μ	Acc_σ	$Angle_\mu$	$Angle_\sigma$	$Disp_\mu$	$Disp_\sigma$
Original values	150.72	0.496	11.49	0.547	1.57	0.278
Multi-objective deterministic optimization	139.21	0.473	10.47	0.563	1.67	0.243
Deterministic optimization verified by FEM	139.94	0.479	10.52	0.571	1.71	0.247
Multi-objective robust optimization	140.86	0.448	10.59	0.551	1.62	0.216
Robust optimization verified by FEM	141.03	0.445	10.65	0.554	1.68	0.223

multi-objective deterministic optimization is better than that obtained by the multi-objective robust optimization, the solutions provided by the latter are more robust. In future work, the uncertainties in both design parameters and material properties will be considered to bring the results of the proposed approach closer to the requirements of engineering applications.

5.2.3.5 Conclusions

(1) A novel NPR energy absorption structure was proposed in this work to improve pedestrian lower leg protection. Comparison with a conventional energy absorption structure composed of the same foam material demonstrated that pedestrian lower leg protection was significantly improved by applying the proposed NPR energy absorption structure to a vehicle front end. The NPR energy absorption structure can be widely used in vehicle front-ends to improve pedestrian protection performance.
(2) Multi-objective optimization of the proposed NPR energy absorption structure with consideration for the perturbation induced by parameter uncertainties was investigated to optimize the design parameters and further enhance the protection performance. A parametric model of the NPR energy absorption structure was first established based the relationships between design parameters to improve the modeling efficiency. Approximation models between inputs and responses were then built by combining the OLHS technique and dual response surface method. Pareto solutions of the NPR energy absorption structure were successfully obtained by applying the MOPSO algorithm and six sigma criteria. A multi-objective deterministic optimization of the NPR energy absorption structure was also conducted in this work to facilitate comparisons. The results showed that the pedestrian lower leg protection of the vehicle front

end was improved significantly by optimizing the NPR energy absorption structure. While the deterministic optimization provided somewhat better protection performance, the proposed robust optimization method provided more robust optimal solutions. This study served as a good example of the safety promotion provided by applying NPR structures with enhanced energy absorption capacity in vehicles.

5.3 Application of Negative Poisson's Ratio Structure in Other Industries

As the negative Poisson's ratio material has special and excellent properties, which makes this material has a broad application prospect, and its possible application areas are.

(1) Using its tensile behavior. It can be used to make shock absorbing materials for rockets, air filters, etc. in the field of aerospace.
(2) Using its compression properties, AUXETIC fibers are used to make composite that are more easily embedded in plastic matrices.
(3) Due to the special characteristics of the pressure can be used to make silencers. And can be used to make their own protective equipment, such as bulletproof undershirts and wrist guards.
(4) Due to its bending characteristics, it can be used in the molding and forming of panels for bodywork, the jet section of locomotives or components in aircraft. For example, the manufacture of fairings.
(5) Production of clothing with high performance quality. Such as thermoplastic underwear, jackets for fighting, armor for body protection, etc. Armor made of this material has better elasticity to adapt to external impact.
(6) To make protective stopping devices on roads.
(7) for making corks. For an ideal bottle stopper manufacturing, it is required to have not only a strict sealing performance. Should also be in the cork after the bottle can be easily pulled out, with rubber materials made of cork in the cork after the bottle easy to form a vacuum with the bottle closed state, cork in difficult to pull out. The use of negative Poisson's ratio of materials made of corks can overcome this defect.
(8) conventional vibration isolators such as rubber vibration isolators can be used in the compression state, but also in the shear state, the internal damping effect is larger, easy aging, heavy load will have greater creep, and mainly for high-frequency vibration isolation. And the ship's auxiliary machine vibration frequency is not high, and the weight is large, rubber vibration isolator in the marine auxiliary machine vibration isolation has certain limitations. Conventional vibration damping impact-resistant structure and materials research so far is difficult to have a major breakthrough, must update the design concept,

special and new structures and materials. The honeycomb structure is widely used in other engineering fields, but no such structure has been applied to the design of vibration isolators. For this reason, this paper designs a macroscopic negative Poisson's ratio honeycomb vibration damping and impact isolator. The calculations show that such vibration isolators have good dynamic performance, and initially reveal the principle of vibration isolation of the new vibration isolators.

(9) Variable-body wing can change the wing shape adaptively according to the change of flight status and environment, thus significantly improving the comprehensive performance of the aircraft and expanding its range of use, which is an important development direction for future advanced aircraft. In the process of continuous wing variation, the wing skin is required to have good in-plane deformation capability and high out-of-plane stiffness to withstand certain aerodynamic loads. Obviously, the traditional hard-shell skin structure cannot meet these requirements at the same time, therefore, the flexible skin structure based on smart materials has been widely studied. The flexible honeycomb structure is applied to the UAV outer wing structure, which increases the area of the wing in the wingspan direction while realizing the deflection of the trailing edge structure, and is of great value to the improvement of the UAV takeoff and landing performance.

(10) The zero Poisson's ratio bullet obtained by introducing negative Poisson's ratio material designed by Mitsubishi Heavy Industries can reduce the lateral expansion generated when the bullet is fired in the barrel, thus improving the bullet motion.

(11) Textiles for apparel Because of the negative Poisson's ratio material has good energy absorption and shape fitting, this kind of special performance fabric has great application in protective clothing and equipment. In some dangerous sports, such as racing, horse racing, ice skating, etc., protective clothing and equipment is essential to protect the wearer from the impact caused by the injury. Protective gear is generally used for elbows, knees and other parts of the body prone to injury (Fig. 5.35), however, most of the protective gear currently on the market is woven from foam, which is less breathable. The three-dimensional negative Poisson's ratio fabric, such as spacer fabric, which is better for human comfort, can be used as a substitute for foam. In addition, because of its isotropic curvature, it can show a rounded arch when bent, and the shape of the human elbow and knee is more closely matched, which can improve its protective performance and the freedom of movement of body parts. This three-dimensional negative Poisson's ratio spacer fabric can be used not only for human body armor but also for shockproof gloves and bulletproof undershirts.

Based on the special needs of some people, negative Poisson's ratio fabric can also be used in daily clothing, such as children's clothing, maternity clothes, bras, etc. In children's clothing, because children's body shape changes quickly, many parents tend to buy loose clothes so as to give their children some room to grow, and loose clothes may be tripped or injured when children are playing.

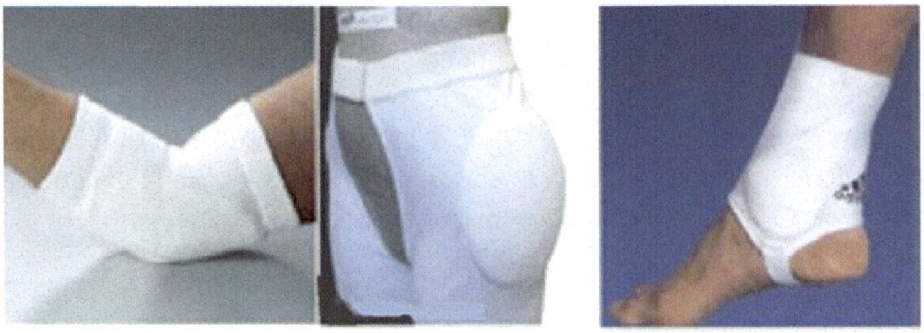

Fig. 5.35 Application of protective clothing and equipment

Negative Poisson than fabric with folding structure is easy to extend both horizontally and vertically, so that the fitted clothes can be worn by children for a longer period of time, thus avoiding frequent purchases and saving costs while ensuring the fit of the clothes. In maternity wear, the current often used in the abdomen or waist elastic fabric, the problem is that when the abdomen is getting bigger, the elastic fabric on its pressure is also bigger, and negative Poisson's ratio fabric can smoothly solve this problem. When the abdomen becomes larger, the negative Poisson's ratio fabric extends in both directions, while naturally forming a rounded arch that fits the shape of the abdomen, so that the abdomen does not need to bear too much pressure (Fig. 5.36). In addition, negative Poisson's ratio fabrics have great application in women's bras. Usually, the requirements for women's bra cups are strong shape fit, good breathability and moisture permeability as well as high wearing comfort, because the negative Poisson's ratio fabric has the same curvature when bending and deforming, its spherical fit is far better than the traditional fabric, and the breathability and moisture permeability of warp knitted interval negative Poisson's ratio fabric is also better than the traditional sponge.

Fig. 5.36 Size Adaptability of Negative Poisson's Ratio Clothing

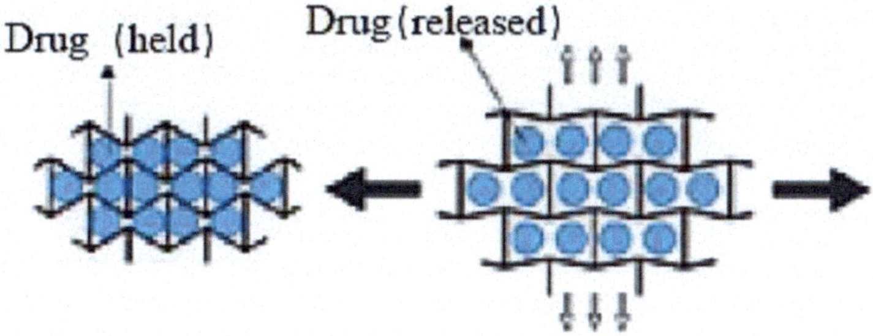

Fig. 5.37 Working principle of intelligent bandage

(12) Medical textiles. In the biomedical field, negative Poisson's ratio fabric can also be used for internal spinal fixators, human-made shaped valves, artificial blood vessels, esophageal stents, etc. Unlike the traditional artificial blood vessels, the artificial blood vessels made of tubular negative Poisson's ratio fabric can be larger in diameter when stretched, which can largely alleviate the danger to human body caused by atherosclerosis, thrombosis and other diseases. Negative Poisson by fabrics can also be made into medical smart bandages, that is, inside the bandage to carry a wound healing agent. When applied to a swollen wound, it opens due to stretching, thus releasing the agent; as the wound gradually heals and the swelling decreases, the bandage contracts and thus stops releasing the agent (Fig. 5.37).

(13) Composite reinforced structure negative Poisson's ratio fiber, yarn or fabric can be used as reinforcement in composite, which can effectively enhance the properties of composite. When the negative Poisson's ratio material is used in the manufacture of sandwich panels, because the bending is expanded outward so as to absorb more energy, rather than the Poisson's ratio of positive material as inward depression is destroyed, therefore, the safety of sandwich panels made of negative Poisson's ratio material can be greatly improved. Due to its good energy absorption, negative Poisson's ratio fabric can be used to produce explosion-proof curtains, which will open up due to the pressure of the shock wave emitted from the explosion, thus capturing the ensuing glass fragments, etc., with explosion-proof function. Coupled with its good spherical fit, the material also has a lot of room for application in helmets, which can mitigate the impact to a certain extent to minimize the damage to the head.

(14) Automotive textiles. The negative Poisson's ratio fabric is used in the car seat belt, when the human body falls forward under the inertia condition, the general seat belt will be narrowed when it is stretched, but the seat belt of the negative Poisson's ratio material will be expanded as a whole, so as to weaken the damage caused by the concentration of the impact pressure on the human body (Fig. 5.38). In addition, this material can also be used for child seat fabric (Fig. 5.39), the current child seat in the use of sponge fabric, if the

use of negative Poisson's ratio material as a child seat fabric and child seat belt, not only can improve its comfort, but also in the safety aspect has greatly improved.

(15) Aerial antenna reflector (patent).

(16) Intelligent hydrostatic bearing based on negative Poisson's ratio structure, using the unique pulling and expanding effect of negative Poisson's ratio structure, can effectively reduce bearing vibration and reduce eccentricity, thus improving motion accuracy, while reducing the reliance on bearing clearance under high-speed conditions, avoiding the traditional high-precision hydrostatic bearing processing difficulties and easy blockage and other problems.

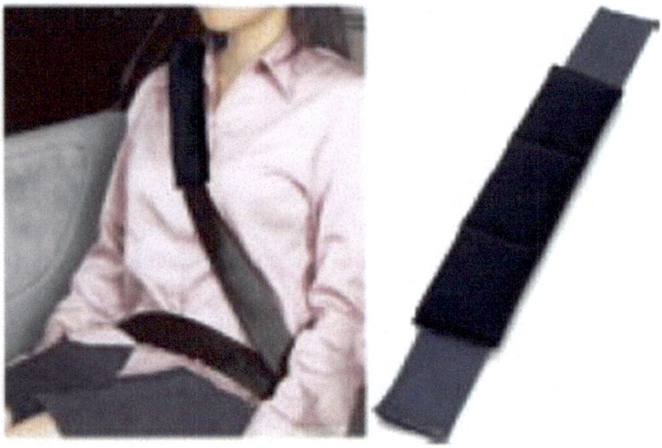

Fig. 5.38 Vehicle safety belt

Fig. 5.39 Child seat fabric

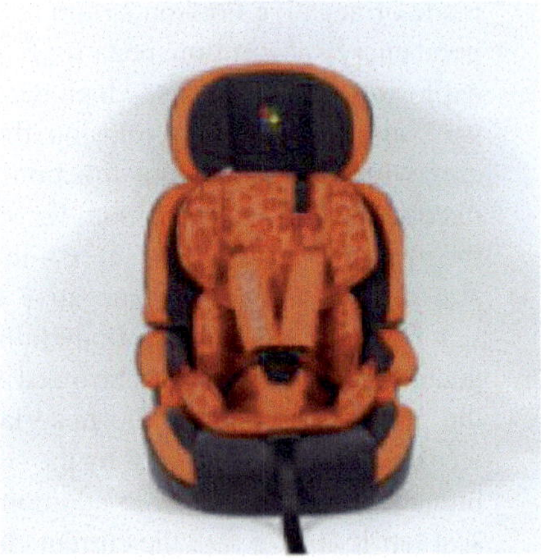

(17) Negative Poisson's ratio materials allow them to be used in sensor design due to their high volume change. A negative Poisson's ratio piezoelectric ceramic sensor designed by the U.S. Navy has dramatically improved the sensitivity of the received signal by up to an order of magnitude compared to the conventional positive Poisson's ratio matrix sensor. For piezoelectric devices, the use of negative Poisson's ratio materials as electrodes can significantly improve their sensitivity.

(18) Negative Poisson's ratio materials also have important applications in biomedicine. In particular, negative Poisson's ratio artificial blood vessels can thicken the vessel wall when internal pressure is generated by blood flow, reducing the possibility of vessel rupture. Negative Poisson's ratio materials can also be used in biomedical applications such as arterial dilators, implant surgery, and suturing.

(19) In 2003, Jones used negative Poisson's ratio foam in his patent application for the invention of headphones This negative Poisson's ratio material makes the shape of the headphones more user-friendly and can be more easily wrapped around different shapes of ear contours, while reducing air leakage due to the special properties of the material, improving the sound quality of the headphones and increasing the comfort level of the ear.

(20) Other fields.

In other fields, negative Poisson's ratio materials with pore structure, such as concave honeycomb materials, where pores unfold in both directions when stretched, can be used in filters; in addition, due to their low bulk modulus, they can be used in sensors, such as hydrophones. In the aerospace field, it can also be used in fan blades of spacecraft engines, thermal protection materials, aircraft forward cone sections, wing sections, etc. Fasteners made of negative Poisson's ratio materials are easier to install and insert than conventional materials, and their radial dimensions can be larger than holes within certain tolerances, and they are subjected to axial pressure during installation so that their diameters also become smaller, thus facilitating installation. After installation to remove the axial pressure, in the axial extension of the radial extension occurs at the same time, which can prevent the fastener from slipping.

5.4 Summary of This Chapter

Negative Poisson's ratio materials, as a material with special mechanical properties, have unique advantages and broad application prospects in many fields due to their special mechanical properties, unique initiating mechanism and lighter structure. In terms of energy absorption, the energy absorption box has an important position in the design of automobile crashworthiness, and the good or bad design of the energy absorption box largely affects the merits of automobile crashworthiness. This chapter introduces a new type of structural energy absorption box-double-arrow-type negative Poisson's ratio structural box, and the analytical results show that the new type of negative Poisson's ratio structural box has a better energy absorption characteristics and specific energy absorption during the collision process, and it can significantly improve the crashworthiness of the automobile. Meanwhile, the microstructural parameters, basis material selection and impact velocity also have significant influence on the performance of double-arrow-type negative Poisson's ratio structural energy-absorbing box, and the structural performance can be enhanced by rationally optimizing the design of the microstructural parameters to meet the different design requirements. In addition, negative Poisson's ratio structures have unique performance advantages in components such as automotive side impact bars and pedestrian protection energy absorbing structures. In addition to the automotive industry, negative Poisson's ratio structures, corresponding to their corresponding application properties, have broad application prospects in the fields of human body protection, aerospace, textile products, biomedicine and national defense.

References

1. Sun, X., Tao, X., Wang, X., Li, J., & Wang, L. (2020). Research on explosion-proof characteristics and optimization design of negative Poisson's ratio honeycomb material. Baozha Yu Chongji/Explosion and Shock Waves, 40(9). https://doi.org/10.11883/bzycj-2020-0011.
2. GARBER, D. P. (1982). TENSILE STRESS-STRAIN BEHAVIOR OF GRAPHITE/EPOXY LAMINATES. NASA Contract Rep.
3. Jones, A., Barthès, J.-P. A., Moulin, C., & Lenne, D. (2014). A rich multi-agent architecture for collaborative multi-touch multi-user devices. Conference Proceedings - IEEE International Conference on Systems, Man and Cybernetics, 2014-January (January), 1107–1112. https://doi.org/10.1109/SMC.2014.6974062.
4. Zhang, Z., Zhang, L., Dong, Y., Chen, H., & Guo, Y. (2023). Mechanical properties of negative Poisson's ratio metamaterial units and honeycomb structures with cosine-like re-entrant structure. Materials Letters, 331. https://doi.org/10.1016/j.matlet.2022.133451.
5. Liu, Y., Ma, D., Ren, J., Zhong, J., & Zhao, C. (2023). Ballistic performance of double-arrow negative Poisson′s ratio structure. Guofang Keji Daxue Xuebao/Journal of National University of Defense Technology, 45(2), 197–207. https://doi.org/10.11887/j.cn.202302023.
6. Çam, S., & Sofuoğlu, H. (2023). An investigation on crashworthiness performance of adhesive bonding in automobile hybrid front bumper system subjected to high-speed impact. International Journal of Impact Engineering, 173. https://doi.org/10.1016/j.ijimpeng.2022.10447
7. Xie, C., Wang, D., Zong, L., Wang, S., & Kong, D. (2023). Multi-objective crashworthiness optimization of energy-absorbing box with gradient lattice structure. Applied Mathematical Modelling, 121, 304–320. https://doi.org/10.1016/j.apm.2023.05.001.
8. Marek, A., Avramidis, K. A., Feuerstein, L., Illy, S., Thumm, M., Wu, C., & Jelonnek, J. (2022). Time-Domain Simulation of Helical Gyro-TWTs With Coupled Modes Method and 3-D Particle Beam. IEEE Transactions on Electron Devices, 69(8), 4546–4552. https://doi.org/10.1109/TED.2022.3182292.
9. Hosseinipour, M., Meymand, S. Z., Craft, M. J., & Ahmadian, M. (2014). Modeling of a Roller Rig for Evaluation of Wheel/Rail Contact Mechanics. 2014 Joint Rail Conference, JRC 2014. https://doi.org/10.1115/JRC2014-3842.
10. An, H., Liu, K., Liang, Z., Qin, M., Huang, Y., & Guo, Z. (2022). Research Review of Object Detection Algorithms in Vehicle Detection. 2022 IEEE International Conference on Electrical Engineering, Big Data and Algorithms, EEBDA 2022, 1337–1341. https://doi.org/10.1109/EEBDA53927.2022.9744735.

11. Patel, M., Khatun, M., Jung, R., & Glab, M. (2021). Simulation-based analysis of highway trajectory planning using high-order polynomial for highly automated driving function. International Conference on Electrical, Computer, Communications and Mechatronics Engineering, ICECCME 2021. https://doi.org/10.1109/ICECCME52200.2021.9591044.
12. Nagel, R. G., Pires, V. F., Silveira, J. L., Cordeiro, A., & Foito, D. (2023). Financial Analysis of Household Photovoltaic Self-Consumption in the Context of the Vehicle-to-Home (V2H) in Portugal. Energies, 16(3). https://doi.org/10.3390/en16031218.
13. Acar, E., & Aplak, H. S. (2016). A model proposal for a multi-objective and multi-criteria vehicle assignment problem: An application for a security organization. Mathematical and Computational Applications, 21(4). https://doi.org/10.3390/mca21040039.
14. Liu, J., Wang, X., Zhai, H., Li, S., Cui, X., & Zhang, Q. (2023). A Method of Vehicle Networking Environment Information Sharing Based on Distributed Fountain Code. SSRN. https://doi.org/10.2139/ssrn.4503736.
15. Li, B., Liu, P., Wu, X., Wang, X., & Zhao, R. (2018). Design and application of agricultural product traceability management platform. Lecture Notes in Computer Science (Including Subseries Lecture Notes in Artificial Intelligence and Lecture Notes in Bioinformatics), 11064 LNCS, 684–693. https://doi.org/10.1007/978-3-030-00009-7_61.
16. Suzuki, K., Tang, K., Alhajyaseen, W., Suzuki, K., & Nakamura, H. (2022). An international comparative study on driving attitudes and behaviors based on questionnaire surveys. IATSS Research, 46(1), 26–35. https://doi.org/10.1016/j.iatssr.2021.10.00.
17. Wei, W., Zhang, F., Xing, Y., Wang, H., & Liu, R. (2023). Research on Mechanical Properties of Origami Aluminum Honeycomb for Automobile Energy Absorbing Box. Materials, 16(1). https://doi.org/10.3390/ma16010141.
18. Zhang, X., & Zhao, X. (2019). Research on structure optimization of automobile energy absorbing box based on collision safety. ACM International Conference Proceeding Series, 200–208. https://doi.org/10.1145/3366194.3366229.
19. Ma, K., Li, B., Yan, Z., & Liu, Q. (2020). Fast Modeling and Analysis of Energy-Absorbing Box. Smart Innovation, Systems and Technologies, 180, 367–375. https://doi.org/10.1007/978-981-15-3867-4_43.
20. Hwang, Y.-H., & Han, J.-H. (2023). Energy absorption optimisation of an origami-shaped crash box under axial loading. International Journal of Crashworthiness. https://doi.org/10.1080/13588265.2023.2196909.
21. Niu, H., Guan, B., Bai, Z. H., Wang, Y., & Zang, Y. (2022). Numerical Simulation and Experimental Study of the Compressive Energy Absorption Characteristics of a Multi-Layered Gradient Egg-Box Structure. Strength of Materials, 54(4), 747–753. https://doi.org/10.1007/s11223-022-00451-x.
22. Chen, X.-K., Li, B.-G., & Lin, Y. (2009). Application of improved RSM in the optimization design of automotive frontal crashworthiness. Beijing Ligong Daxue Xuebao/Transaction of Beijing Institute of Technology, 29(12), 1076–1079+1084.
23. Pandey, A. K., Kumar, A., Tripathi, J. P., & Ehtesham Hasan, Md. (2023). Analysis of a high-speed low-torque variable displacement hydro-motor efficiency at different displacement conditions. Materials Today: Proceedings, 79, 247–249. https://doi.org/10.1016/j.matpr.2022.11.114,161.

The manufacturer's authorised representative in the EU is Springer Nature Customer Service Centre GmbH, Europaplatz 3, 69115 Heidelberg, Germany. If you have any concerns regarding our products, please contact ProductSafety@springernature.com

Printed and bound by CPI Group (UK) Ltd, Croydon, CR0 4YY
01/10/2026
02225945-0002